ELECTRONICS FOR ELECTRICIANS

6TH EDITION

ELECTRONICS FOR
ELECTRICIANS

6TH EDITION

Stephen L. Herman

DELMAR
CENGAGE Learning™

Australia • Brazil • Japan • Korea • Mexico • Singapore • Spain • United Kingdom • United States

DELMAR
CENGAGE Learning™

Electronics for Electricians, 6E
Stephen L. Herman

Vice President, Editorial: Dave Garza

Director of Learning Solutions: Sandy Clark

Acquisitions Editor: Stacy Masucci

Managing Editor: Larry Main

Senior Product Manager: John Fisher

Editorial Assistant: Andrea Timpano

Vice President, Marketing: Jennifer Baker

Marketing Director: Deborah Yarnell

Marketing Manager: Katie Hall

Associate Marketing Manager: Jillian Borden

Senior Production Director: Wendy Troeger

Production Manager: Mark Bernard

Content Project Manager: Barbara LeFleur

Senior Art Director: David Arsenault

Technology Project Manager: Joe Pliss

For product information and technology assistance, contact us at
Cengage Learning Customer & Sales Support, 1-800-354-9706

For permission to use material from this text or product,
submit all requests online at **www.cengage.com/permissions**.
Further permissions questions can be e-mailed to
permissionrequest@cengage.com

Library of Congress Control Number: 2011922301

ISBN-13: 978-1-1111-2780-0

ISBN-10: 1-1111-2780-8

Delmar
5 Maxwell Drive
Clifton Park, NY 12065-2919
USA

Cengage Learning is a leading provider of customized learning solutions with office locations around the globe, including Singapore, the United Kingdom, Australia, Mexico, Brazil, and Japan. Locate your local office at:
international.cengage.com/region

Cengage Learning products are represented in Canada by
Nelson Education, Ltd.

To learn more about Delmar, visit **www.cengage.com/delmar**

Purchase any of our products at your local college store or at our preferred online store **www.cengagebrain.com**

Notice to the Reader

Publisher does not warrant or guarantee any of the products described herein or perform any independent analysis in connection with any of the product information contained herein. Publisher does not assume, and expressly disclaims, any obligation to obtain and include information other than that provided to it by the manufacturer. The reader is expressly warned to consider and adopt all safety precautions that might be indicated by the activities described herein and to avoid all potential hazards. By following the instructions contained herein, the reader willingly assumes all risks in connection with such instructions. The publisher makes no representations or warranties of any kind, including but not limited to, the warranties of fitness for particular purpose or merchantability, nor are any such representations implied with respect to the material set forth herein, and the publisher takes no responsibility with respect to such material. The publisher shall not be liable for any special, consequential, or exemplary damages resulting, in whole or part, from the readers' use of, or reliance upon, this material.

Printed in the United States of America
2 3 4 5 6 7 14 13 12 11

TABLE OF CONTENTS

PREFACE

Electronics for Electricians, Sixth Edition, is intended for students in the electrical field. The text assumes that the student has knowledge of basic electricity and basic circuits. *Electronics for Electricians* is a hands-on approach to the subject of electronics. Components are presented in a straightforward practical manner as opposed to being merely mathematical concepts. Components are explained from a standpoint of how they operate and how they can be used in a circuit. Laboratory experiments use common devices that are readily available from a number of sources. The laboratory experiments have been student tested and help the student to learn by doing.

The sixth edition of *Electronics for Electricians* has added coverage on the following:

- Extended coverage of light-emitting diodes
- Updated illustrations
- Foreword concerning the connection of circuits in the laboratory
- Corrections from previous editions

The sixth edition of *Electronics for Electricians* is an effective text for teaching electrical students electronics from a practical hands-on approach. In the back of the textbook are hands-on laboratory experiments. Appendix C contains a parts list of the necessary components, common to the industry, required to connect all the experiments. However, any component with similar characteristics can be used if the listed component is not available. For example, two of the transistors listed in Appendix C are the 2N2222 and the 2N2907. Other transistors may be substituted if they have similar voltage and current ratings.

SUPPLEMENTS TO THIS BOOK

Also available to the instructor is the *Instructor Resources to Accompany Electronics for Electricians*, Sixth Edition. Thoroughly updated to reflect changes to the sixth edition book, the *Instructor Resources* contain:

- Instructor's Guide with answers to the text's Review Questions
- PowerPoint presentations
- Test bank

(Order #: 1111127816)

Visit us now at our newly designed web site, www.cengage.com/community/electrical, featuring author biographies, discussion board, electrical blogs, industry articles, and more.

ABOUT THE AUTHOR

Stephen L. Herman has been both a teacher of industrial electricity and a working industrial electrician. He attended Stephen F. Austin University in Nacogdoches, Texas, and Catawba Valley Technical College in Hickory, North Carolina. He has worked as a maintenance electrician for Superior Cable Corp. and National Pipe and Tube Corp. Mr. Herman was employed at Randolph Technical College in Asheboro, North Carolina, for nine years as the Electrical Installation and Maintenance Instructor. During that time, he served as president for the North Carolina Electricals Instructors Association for six years. Mr. Herman was lead instructor for the Electrical Technology Curriculum at Lee College in Baytown, Texas, for 20 years. During that time, he received the Halliburton Education Foundation Award for Excellence in Teaching. He is now retired from teaching but continues to update and write textbooks. He and his wife, Debbie, now reside in Pittsburg, Texas.

ACKNOWLEDGMENTS

The author and Delmar Cengage Learning would like to acknowledge and thank the review panel for their suggestions and comments. Thanks go to:

Louis A. Gilstrap, Vatterott College, Kansas City, Missouri

Kenneth R. Ludington, Guilford Technical Community College, Jamestown, North Carolina

Sebastian Mark Rose, Calhoun Community College, Tanner, Alabama

Ron Taylor Calhoun Community College, Decatur, Alabama

UNIT 1

SEMICONDUCTORS
AND RESISTORS

OBJECTIVES

After studying this unit, the student should be able to:

- Discuss the differences among the atomic structure of conductors, insulators, and semiconductors.

- Give an explanation of how P- and N-type materials are made.

- Describe a lattice structure.

- Discuss the conduction of electric current through semiconductor materials.

SEMICONDUCTOR MATERIAL

Solid-state electronic devices are so named because they have no moving parts to wear. They are constructed from solid pieces of material called **semiconductors**. Although there are some vacuum- and gas-filled tubes still used in industry, today most electronic devices are solid state. Solid-state devices have the advantage of being less susceptible to shock and vibration; they can operate at lower voltages, are much smaller in size, waste less power due to heat, and generally have a much longer life span.

Semiconductors differ from insulators or conductors in their atomic structure. Conductors are materials that conduct electricity easily. They are generally made from materials that have one, two, or three **valence electrons**, Figure 1-1. Valence electrons are electrons located in the outer or valence shell of an atom. Silver is the best natural conductor of electricity, followed by copper, gold, and aluminum. Silver, copper, and gold have one valence electron. Aluminum contains three valence electrons. Although many elements contain few valence electrons, they are not as good conductors of electricity as are silver, copper, gold, or aluminum. A table of elements showing the number of valence electrons is shown in Figure 1-2.

FIGURE 1-1 *Conductors contain few valence electrons.*

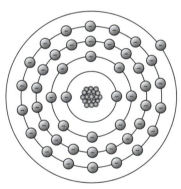

Silver Atom

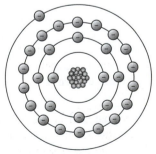

Copper Atom

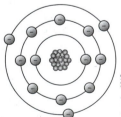

Aluminum Atom

© Cengage Learning 2012

ATOMIC NUMBER	NAME	VALENCE ELECTRONS	SYMBOL	ATOMIC NUMBER	NAME	VALENCE ELECTRONS	SYMBOL	ATOMIC NUMBER	NAME	VALENCE ELECTRONS	SYMBOL
1	Hydrogen	1	H	37	Rubidium	1	Rb	73	Tantalum	2	Ta
2	Helium	2	He	38	Strontium	2	Sr	74	Tungsten	2	W
3	Lithium	1	Li	39	Yttrium	2	Y	75	Rhenium	2	Re
4	Berylium	2	Be	40	Zirconium	2	Zr	76	Osmium	2	Os
5	Boron	3	B	41	Hiobium	1	Nb	77	Iridum	2	Ir
6	Carbon	4	C	42	Molybdenum	1	Mo	78	Platinum	1	Pt
7	Nitrogen	5	N	43	Technetium	2	Tc	79	Gold	1	Au
8	Oxygen	6	O	44	Ruthenium	1	Ru	80	Mercury	2	Hg
9	Fluoride	7	F	45	Rhodium	1	Rh	81	Thallium	3	Tl
10	Neon	8	Ne	46	Palladium	-	Pd	82	Lead	4	Pb
11	Sodium	1	Na	47	Silver	1	Ag	83	Bismuth	5	Bi
12	Magnesium	2	Ma	48	Cadmium	2	Cd	84	Polonium	6	Po
13	Aluminum	3	Al	49	Indium	3	In	85	Astatine	7	At
14	Silicon	4	Si	50	Tin	4	Sn	86	Radon	8	Rd
15	Phosphorus	5	P	51	Antimony	5	Sb	87	Francium	1	Fr
16	Sulfur	6	S	52	Tellurium	6	Te	88	Radium	2	Ra
17	Chlorine	7	Cl	53	Iodine	7	I	89	Actinium	2	Ac
18	Argon	8	A	54	Xenon	8	Xe	90	Thorium	2	Th
19	Potassium	1	K	55	Cesium	1	Cs	91	Protactinium	2	Pa
20	Calcium	2	Ca	56	Barium	2	Ba	92	Uranium	2	U
21	Scandium	2	Sc	57	Lanthanum	2	La				
22	Titanium	2	Ti	58	Cerium	2	Ce				
23	Vanadium	2	V	59	Praseodymium	2	Pr		**ARTIFICIAL ELEMENTS**		
24	Chromium	1	Cr	60	Neodymium	2	Nd				
25	Manganese	2	Mn	61	Promethium	2	Pm	93	Neptunium	2	Np
26	Iron	2	Fe	62	Samarium	2	Sm	94	Plutonium	2	Pu
27	Cobalt	2	Co	63	Europium	2	Eu	95	Americium	2	Am
28	Nickel	2	Ni	64	Gadolinium	2	Gd	96	Curium	2	Cm
29	Copper	1	Cu	65	Tebrium	2	Tb	97	Berkelium	2	Bk
30	Zinc	2	Zn	66	Dysprosium	2	Dy	98	Californium	2	Cf
31	Gallium	3	Ga	67	Holmium	2	Ho	99	Einsteinium	2	E
32	Germanium	4	Ge	68	Erbium	2	Er	100	Fermium	2	Fm
33	Arsenic	5	As	69	Thulium	2	Tm	101	Mendelevium	2	Mv
34	Selenium	6	Se	70	Ytterbium	2	Yb	102	Nobelium	2	No
35	Bromine	7	Br	71	Lutetium	2	Lu	103	Lawrencium	2	Lw
36	Krypton	8	Kr	72	Hafnium	2	Hf				

FIGURE 1-2 *Table of elements*

Semiconductors are materials that contain four valence electrons, such as carbon, silicon, germanium, tin, and lead. Silicon and germanium are the materials commonly used in the production of semiconductor devices, Figure 1-3. Silicon is the most common element used in the production of semiconductor devices because it has the ability to withstand more heat than germanium.

Regardless of which semiconductor material is used, it should be intrinsic (without impurities). When the atoms of a pure semiconductor material combine, the four valence electrons of each atom bond with the orbits of neighboring atoms to form a crystal

FIGURE 1-3 *The most common semiconductor materials are germanium and silicon. Each contain four valence electrons.*

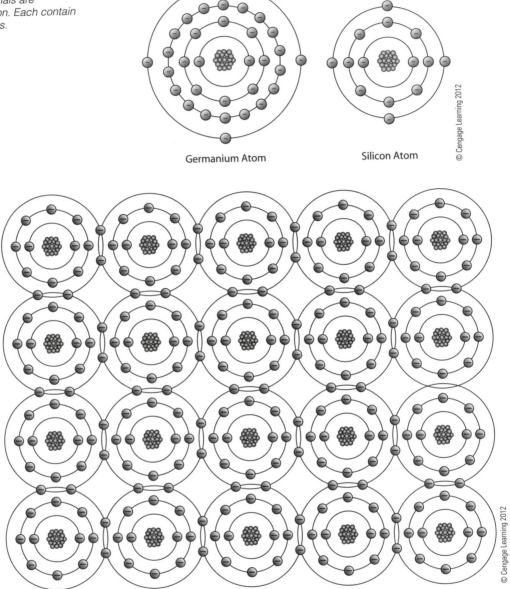

Germanium Atom   Silicon Atom

© Cengage Learning 2012

FIGURE 1-4 *Semiconductor atoms join together to form a covalence bond.*

© Cengage Learning 2012

structure called a **lattice structure**. This crystal structure is made possible because of a covalence bonding, Figure 1-4. Another common semiconductor material that forms a crystal structure due to covalence bonding is carbon. Carbon forms a crystal structure generally referred to as diamond.

Intrinsic silicon and germanium are neither good conductors nor insulators in their natural state. The resistance of either material is greatly affected by temperature. Although most conductive materials, such as copper or aluminum, increase their resistance with an increase in temperature, semiconductive materials decrease their resistance with an increase in temperature. Although both silicon and germanium are semiconductor materials, intrinsic germanium exhibits about 1/1000th the amount of resistance as intrinsic silicon. Silicon is the most used semiconductor material, however, because of its ability to withstand heat. It is the addition of impurities in the lattice structure that makes intrinsic semiconductor material useful for the conduction of electric current.

N-type Materials

The process used to add impurities, or elements with a different number of valence electrons, to an intrinsic semiconductor material is called **doping**. Doping is a highly controlled process that involves vaporizing certain elements in a high-temperature oven with disks of intrinsic semiconductor material. N-type material is formed by doping the semiconductor material with a pentavalent element such as arsenic, antimony, or phosphorus. Pentavalent elements contain five valence electrons. The doped material is called an extrinsic semiconductor material. The doping process causes a pentavalent atom to combine with the atoms of the semiconductor lattice structure, Figure 1-5. Four of the valence electrons combine with the semiconductor atoms to form the covalence bond, but the fifth electron is very loosely held and is free to produce current flow. Because the extrinsic semiconductor material now contains extra electrons and electrons are negative particles, it is referred to as N (negative)-type material.

P-type Material

The production of P-type material is very similar to the production of N-type material, except that a trivalent impurity such as aluminum, boron, gallium, or indium is used. Trivalent elements contain three valence electrons. When the trivalent element is doped to the semiconductor material, the three valence electrons form a covalent bond with the other semiconductor atoms, Figure 1-6. Because the impurity contains three valence electrons instead of four, a hole is left in the lattice structure when it bonds with the semiconductor atoms. These holes are spaces left in the material that can accept an electron. Because the hole represents a missing electron and an electron has a negative charge, the hole is a missing negative charge, which makes it positive. Holes are considered to be positive current carriers. Because this type of material has a net positive charge, it is referred to as a P (positive)-type material.

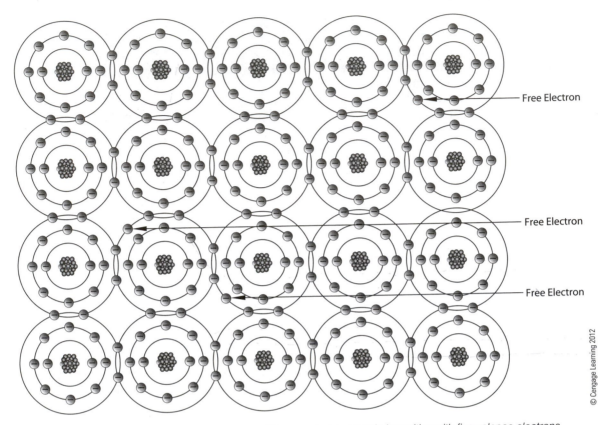

Free Electron

Free Electron

Free Electron

© Cengage Learning 2012

FIGURE 1-5 *N-type materials contain impurities with five valence electrons.*

Conduction Through Semiconductors

The majority current carrier in an N-type material is electrons because N-type material contains extra free electrons. The majority current carrier in a P-type material is considered to be holes because P-type material contains a lack of electrons or an excess of holes. In a P-type material, conduction is caused by hole flow. When an electron leaves the valence shell of an atom, it leaves a hole. When the electron comes in proximity of a hole, it fills it. Holes are continually being created and filled by the moving electrons. Imagine a brick wall with several bricks missing, Figure 1-7. Assume the bricks to be electrons and the missing bricks to be holes. If a brick were to move to the left and fill a hole, a hole would appear in the position the brick was formerly occupying. If this process continued in rapid succession, it would appear that bricks were moving to the left and holes were moving to the right. This same basic process takes place inside the semiconductor material when current flows through it.

Hole

Hole

Hole

FIGURE 1-6 *P-type materials contain impurities with three valence electrons.*

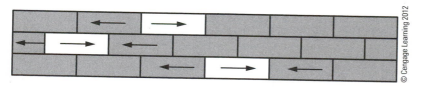

FIGURE 1-7 *Bricks move to the left and holes appear to move to the right.*

Joining P-type and N-type Materials

P- and N-type materials by themselves are not very useful. When these materials become joined, however, literally thousands of electronic devices can be created. When a piece of P-type and a piece of N-type material are joined, a barrier potential or barrier voltage is formed. The barrier voltage and the ability of P- and N-type materials to be joined at all is a result of fixed ion charge in the extrinsic semiconductor material.

To understand fixed ion charges, it is necessary to understand two general principles concerning ions:

1. Ions are charged atoms. A neutral atom has the same number of protons in the nucleus as it has electrons in orbit. The positive charges of the protons are balanced by the negative charges of the electrons, producing a net charge of zero for a balanced atom. If an atom should lose one or more electrons, there will be more positive charges in the nucleus than there are negative electrons. The atom now has a net positive charge and is a positive ion.

   If an atom should gain one or more electrons, it will have more negative electrons than positive protons, resulting in a net negative charge. The atom is now a negative ion.

2. Because ions are atoms, they have a nucleus. If these ions are part of a solid material, they are relatively immobile and not free to move around.

When a piece of P-type and a piece of N-type material are joined together, Figure 1-8, the positive holes in the P-type material are attracted to the free negative electrons in the N-type material. Some of the positive holes are attracted to the negative electrons in the N-type material and drift toward the junction. Some of the electrons are attracted to the positive holes in the P-type material and drift toward the junction. These oppositely charged particles combine to form an electron-hole pair that has no charge.

Note in Figure 1-8 that positive ions are formed at the edge of the N-type material and negative ions are formed at the edge of the

FIGURE 1-8 *Electrons drift across the junction into the P region and holes drift across the junction into the N region.*

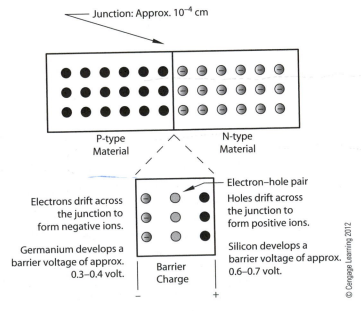

© Cengage Learning 2012

P-type material. This is a result of the doping process that formed the P- and N-type materials. The ions are stationary; consequently, when positive holes combine with the negative electrons, some of the positive ions lose positive charges and become negative ions, and some of the negative ions lose negative charges and become positive ions. It is the formation of this barrier charge that prevents additional holes and electrons from drifting across the junction and causing both the P- and N-type materials from becoming electrically neutral.

The barrier potential is internal to the device and cannot be measured directly. Measurement is made by applying an external voltage and breaking down the barrier to permit current to flow across the junction. Germanium devices require a voltage of 0.3 to 0.4 volt to overcome the barrier potential, and silicon devices require a voltage of 0.6 to 0.7 volt. Silicon requires a higher voltage to overcome the barrier potential because of greater stability in the covalence bonding due to its lower atomic number.

Temperature Effects on the Junction

The barrier potential values of 0.3 to 0.4 volt for germanium and 0.6 to 0.7 volt for silicon are based on an ambient temperature of 25°C. If the temperature is increased, the additional thermal energy produces more minority current carriers and the barrier potential decreases. If the ambient temperature is decreased, the barrier potential will increase. PN junctions can be used as very accurate temperature sensors by permitting a constant amount of current to flow through them and measuring the voltage drop across the device, Figure 1-9. The amount of voltage drop across the pn junction will be proportional to its temperature. An increase in temperature will cause a corresponding decrease in the voltage drop.

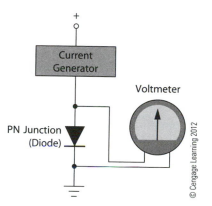

FIGURE 1-9 *Measuring temperature by measuring the voltage drop across a PN junction*

ELECTRICAL RATINGS OF SEMICONDUCTOR DEVICES

Semiconductor devices can have several ratings, such as voltage, current, power, and frequency. The voltage rating indicates the amount of voltage a component can withstand in the reverse direction without breaking down. Voltage ratings of 50 volts, 100 volts, 200 volts, 400 volts, and 1000 volts are common. It should be noted that these voltage ratings are given as PIV (peak *inverse* voltage) or PRV (peak *reverse* voltage) values. The words inverse and reverse both indicate that the value is for voltage applied in the reverse direction. Most semiconductor devices exhibit a very low voltage drop when they conduct. The word "peak" indicates that it is a peak voltage value. Most semiconductor devices are intended to operate on direct current. When they are used in an alternating current circuit, the rms value of voltage must be converted to the peak value to determine whether the device is being used within its rating. The peak value can be determined by multiplying the rms value by 1.414.

EXAMPLE: A diode is connected into a 240-volt AC system. What minimum PIV or PRV rating should the diode have?

$$Peak = rms \times 1.414$$

$$Peak = 240 \times 1.414$$

$$Peak = 339.36 \text{ volts}$$

The diode in this example should have a minimum voltage rating of 339.36 volts.

Current Rating

The current rating indicates the maximum amount of current the device can control. Depending on the device, it may not be able to control the listed current if it exhibits a large voltage drop while conducting current. Some devices such as **diodes, SCRs (silicon-controlled rectifiers)**, and **triacs** are designed to operate completely turned on and can generally handle the rated current. **Transistors**, however, can have large voltage drops and may not be able to control the maximum rated current.

All solid-state devices are made from a combination of P- and N-type materials. The type of device formed is determined by how the P- and N-type materials joined together. The number of layers of material and the thickness of various layers play an important part in determining what type of device will be formed. The diode is often called a pn junction because it is made by joining together a piece of P-type and a piece of N-type material, Figure 1-10. The transistor is made by joining three layers of semiconductor material together, Figure 1-11. Regardless of the type of solid-state device being used, it is made by joining together P- and N-type materials.

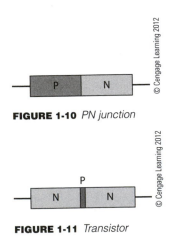

FIGURE 1-10 *PN junction*

FIGURE 1-11 *Transistor*

© Cengage Learning 2012

RESISTORS

Another component found in almost every electronic circuit is the resistor. Although the resistor in not a semiconductor device, it is just as important to the operation of electronic circuits. Resistors are commonly used to perform two functions in a circuit. One is to limit the flow of current through the circuit. In Figure 1-12, a 30-ohm resistor is connected to a 15-volt battery. The current in this circuit is limited to a value of 0.5 amp.

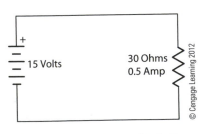

FIGURE 1-12 *Resistor used to limit the flow of current*

© Cengage Learning 2012

$$I = \frac{E}{R}$$

$$I = \frac{15}{30}$$

$$I = 0.5 \text{ Amp}$$

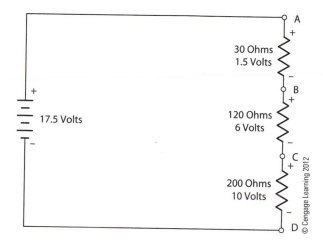

FIGURE 1-13 *Resistors used as a voltage divider*

If this resistor were not present, the circuit current would be limited only by the resistance of the conductor, which would be very low, causing a large amount of current to flow. Assume, for example, that the wire has a resistance of 0.0001 ohm. When the wire is connected across the 15-volt power source, a current of 150,000 amps would try to flow through the circuit (15/0.0001 = 150,000). This is commonly known as a short circuit.

The second principal function of resistors is to produce a voltage divider, as shown in Figure 1-13. These three resistors are connected in series with a 17.5-volt battery. A voltmeter placed between different points in the circuit would indicate the voltages shown below.

A–B	1.5 volts
A–C	7.5 volts
A–D	17.5 volts
B–C	6 volts
B–D	16 volts
C–D	10 volts

By connecting resistors of the proper value, almost any voltage desired can be obtained. Voltage dividers were used to a large extent in vacuum tube circuits many years ago. Voltage divider circuits are still used today in applications involving field effect transistors (FETs) and in multirange voltmeter circuits.

Fixed Resistors

Fixed resistors have only one ohmic value, which cannot be changed or adjusted. There are several different types of fixed resistors. One of the most common types of fixed resistors is the composition carbon resistor. Carbon resistors are made from a compound of carbon graphite and a resin bonding material. The proportions of carbon and resin

FIGURE 1-14 *Compound carbon resistor*

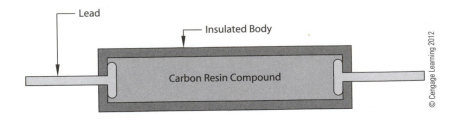

Lead

Insulated Body

Carbon Resin Compound

© Cengage Learning 2012

material determine the value of resistance. This compound is enclosed in a case of nonconductive material with connecting leads, Figure 1-14.

Carbon resistors are very popular for most applications because they are inexpensive and readily available. They are made in standard values that range from about 1 ohm to about 22 million ohms, and they can be obtained in power ratings of ⅛, ¼, ½, 1, and 2 watts. The power rating of the resistor is indicated by its size. A ½-watt resistor is approximately ⅜ inch in length and ⅛ inch in diameter. A 2-watt resistor has a length of approximately ¹¹⁄₁₆ inch and a diameter of approximately ⁵⁄₁₆ inch, Figure 1-15. The 2-watt resistor is larger than the ½ watt or 1 watt, because it must have a larger surface area to be able to dissipate more heat. Examples of 2-watt, 1-watt, and ½-watt resistors are shown in Figure 1-16. Although carbon resistors have a lot of desirable characteristics, they have one characteristic that is not desirable. Carbon resistors will change their value with age or if they are overheated. Carbon resistors generally increase rather than decrease in value.

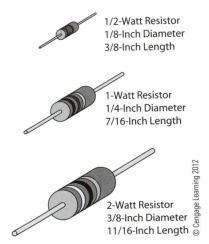

1/2-Watt Resistor
1/8-Inch Diameter
3/8-Inch Length

1-Watt Resistor
1/4-Inch Diameter
7/16-Inch Length

2-Watt Resistor
3/8-Inch Diameter
11/16-Inch Length

© Cengage Learning 2012

FIGURE 1-15 *Power rating is indicated by size.*

Metal Film Resistors

Another type of fixed resistor is the metal film resistor. Metal film resistors are constructed by applying a film of metal to a ceramic rod in a vacuum, Figure 1-17. The resistance is determined by the type of

FIGURE 1-16 *Composition carbon resistors*

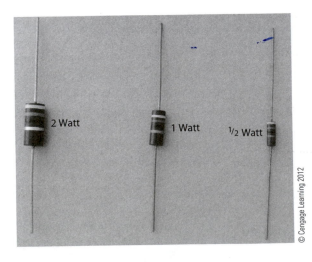

2 Watt   1 Watt   ½ Watt

© Cengage Learning 2012

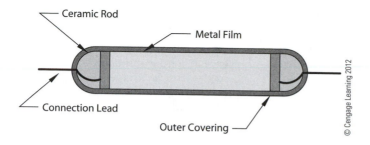

FIGURE 1-17 *Construction of a metal film resistor*

metal used to form the film and by the thickness of the film. Typical thickness for the film is from 0.00001 to 0.00000001 inch. Leads are then attached to the film coating, and the entire assembly is covered with a coating. These resistors are superior to carbon resistors in several respects. Metal film resistors do not change their value with age, and their tolerance is generally better than carbon resistors. Carbon resistors commonly have a tolerance range of 20 percent, 10 percent, or 5 percent. Metal film resistors generally range in tolerance from 2 percent to 0.1 percent. The disadvantage of the metal film resistor is that it is higher in cost.

Carbon Film Resistor

Another type of resistor that is constructed in a similar manner is the carbon film resistor. This resistor is made by coating a ceramic rod with a film of carbon instead of metal. Carbon film resistors are less expensive to manufacture than metal film resistors and can have a better tolerance rating than composition carbon resistors. Due to improved manufacturing techniques, carbon film resistors are rapidly replacing composition carbon resistors. Carbon film resistors do not suffer from the tendency of increased resistance value with age, as do the composition carbon resistors.

Wire Wound Resistors

Wire wound resistors are made by winding a piece of resistive wire around a ceramic core, Figure 1-18. The resistance of a wire wound resistor is determined by three factors:

- The type of material used to make the resistive wire
- The diameter of the wire
- The length of the wire

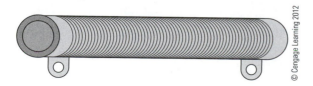

FIGURE 1-18 *Wire wound resistor*

FIGURE 1-19 *Typical wire wound resistor*

Wire wound resistors can be found in various case styles and sizes. These resistors are generally used when a high power rating is needed. Wire wound resistors can operate at higher temperatures than any other type of resistor. A wire wound resistor that has a hollow center is shown in Figure 1-19. The disadvantage of wire wound resistors is they are expensive and generally require a large amount of space for mounting. They can also exhibit an amount of inductance in circuits that operate at high frequencies. This added inductance can cause problems to the rest of the circuit. Inductance will be covered in later units of this text.

Color Code

The values of a resistor can often be determined by the color code. Many resistors have bands of color that are used to determine the resistance value, tolerance, and in some cases reliability. The color bands represent numbers. Each color represents a different numerical value. The chart shown in Figure 1-20 lists the most commonly used colors and the number values assigned to them. The resistor shown beside the color chart illustrates how to determine the value of a resistor. Resistors can have from three to five bands of color. Resistors that have a tolerance of ±20 percent have only three color bands. Most resistors contain four bands of color. For resistors with tolerances that range from ±10 percent to ±2 percent, the first two color bands represent number values. The third color band is called the multiplier. This means to multiply the first two numbers by 10, the number of times indicated by the value of the third band. The fourth band indicates the tolerance. For example, assume a resistor has color bands of brown, green, red, and silver. The first two bands represent

FIGURE 1-20 *Resistor color code chart*

Resistor Color Code		
Color	Number	Tolerance
Black	0	
Brown	1	1%
Red	2	2%
Orange	3	
Yellow	4	
Green	5	
Blue	6	
Violet	7	
Gray	8	
White	9	
Gold	(0.1 multiplier)	5%
Silver	(0.01 multiplier)	10%
No color		20%

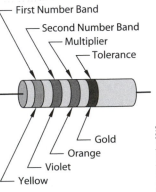

First Number Band
Second Number Band
Multiplier
Tolerance
Gold
Orange
Violet
Yellow

the numbers 1 and 5 (brown is 1 and green is 5). The third band is red, which has a number value of 2. The number 15 should be multiplied by 10 two times. The value of the resistor is 1500 ohms. Another method that is simpler to understand is to add the number of zeros to the first two numbers indicated by the multiplier band. The multiplier band in this example is red, which has a numeric value of 2. Add two zeros to the first two numbers. The number 15 becomes 1500.

The fourth band is the tolerance band. The tolerance band in this example is silver, which means ±10 percent. This resistor should be 1500 ohms plus or minus 10 percent. To determine the value limits of this resistor, find 10 percent of 1500.

$$1500 \times 0.10 = 150$$

The value can range from 1500 + 10 percent or 1500 + 150 = 1650 ohms to 1500 − 10 percent or 1500 − 150 = 1350 ohms. An example of this resistor is shown in the back of the textbook.

Resistors that have a tolerance of ±1 percent and some military resistors contain five bands of color.

EXAMPLE 1: Assume a resistor contains the following bands of color:

> First band = Brown
> Second band = Black
> Third band = Black
> Fourth band = Brown
> Fifth band = Brown

The brown fifth band indicates that this resistor has a tolerance of ±1 percent. To determine the value of a 1 percent resistor, the first three bands are numbers and the fourth band is the multiplier. In this example, the first band is brown, which has a number value of 1. The next two bands are black, which represent a number value of 0. The fourth band is brown, which means add one 0 to the first three numbers. The value of this resistor is 1000 ohms ±1 percent. An example of this resistor is shown in the back of the textbook.

EXAMPLE 2: A five-band resistor has the following color bands:

> First band = Red
> Second band = Orange
> Third band = Violet
> Fourth band = Red
> Fifth band = Brown

The first three bands represent number values. Red is 2, orange is 3, and violet is 7. The fourth band is the multiplier, in this case red represents 2. Add two 0s to the number 237. The value of the resistor is 23,700 ohms. The fifth band is brown, which indicates a tolerance of ±1 percent.

Military resistors often have five bands of color also. These resistors are read in the same manner as a resistor with four bands of color. The fifth band can represent different things. A fifth band of orange or yellow is used to indicate reliability. Resistors with a fifth band of orange have a reliability good enough to be used in missile systems, and resistors with a fifth band of yellow can be used in space flight equipment. A military resistor with a fifth band of white indicates the resistor has solderable leads.

Resistors with tolerance ratings ranging from 0.5 percent to 0.1 percent will generally have their values printed directly on the resistor.

Gold and Silver as Multipliers

The colors gold and silver are generally found in the fourth band of a resistor, but they can be used in the multiplier band also. When the color gold is used as the multiplier band, it means the first two numbers are divided by 10. If silver is used as the multiplier band, it means the first two numbers are divided by 100. For example, assume a resistor has color bands of orange, white, gold, and gold. The value of this resistor is 3.9 ohms with a tolerance of ± 5 percent (orange = 3, white = 9, gold means to divide 39 by 10, which = 3.9, and gold in the forth band means ± 5 percent tolerance).

Standard Resistance Values

Fixed resistors are generally produced in standard values. The higher the tolerance value, the fewer resistance values available. Standard resistor values are listed in the chart shown in the back of the textbook. In the column under 10 percent, there are only 12 values of resistors listed. These standard values, however, can be multiplied by factors of 10. Notice that one of the standard values listed is 33 ohms. There are also standard values in 10 percent resistors of 0.33, 3.3, 330, 3300, 33,000, 330,000, and 3,300,000 ohms. The 5 percent column shows there are 24 resistor values and the 1 percent column lists 96 values. All of the values listed in the chart can be multiplied by factors of 10 to obtain other resistance values. A full color code chart for standard value resistors is shown in the back of the textbook.

Variable Resistors

A variable resistor is a resistor whose values can be changed or varied over a range. Variable resistors can be obtained in different case styles and power ratings. Figure 1-21 illustrates how a variable resistor is constructed. In this example, a resistive wire is wound in a circular pattern, and a sliding tap makes contact with the wire. The value of resistance can be adjusted between one end of the resistive wire and

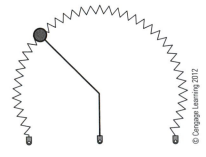

FIGURE 1-21 *Variable resistor*

© Cengage Learning 2012

the sliding tap. If the resistive wire has a total value of 100 ohms, the resistor can be set between the values of 0 and 100 ohms.

A variable resistor with three terminals is shown in Figure 1-22. This type of resistor has a wiper arm inside the case that makes contact with the resistive element. The full resistance value is between the two outside terminals, and the wiper arm is connected to the center terminal. The resistance between the center terminal and either of the two outside terminals can be adjusted by turning the shaft and changing the position of the wiper arm. Wire wound variable resistors of this type can be obtained also. The advantage of the wire wound type is a higher power rating.

FIGURE 1-22 *A three-terminal variable resistor*

The resistor shown in Figure 1-22 can be adjusted from its minimum to maximum value by turning the control approximately ¼ of a turn. In some types of electrical equipment, this range of adjustment may be too coarse to allow for sensitive adjustments. When this becomes a problem, a multiturn resistor can be used, Figure 1-23. Multiturn variable resistors operate by moving the wiper arm with a screw of some number of turns. They generally range from 3 turns to 10 turns. If a 10-turn variable resistor is used, it will require 10 turns of the control knob to move the wiper from one end of the resistor to the other, instead of ¼ of a turn.

Variable Resistor Terminology

Variable resistors are known by several common names. The most popular name is *pot*, which is shortened from the word "potentiometer." Another common name is rheostat. A rheostat is actually a variable resistor that has only two terminals instead of three, but three-terminal variable resistors are often referred to as rheostats also. A potentiometer describes how a variable resistor is used rather than some specific type of resistor. The word "potentiometer" comes from the word "potential" or voltage. A potentiometer is a variable resistor used to provide a variable voltage, as shown in Figure 1-24. In this

FIGURE 1-23 *Multiturn variable resistor*

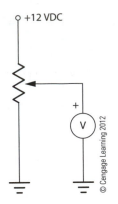

FIGURE 1-24 *Variable resistor used as a potentiometer*

example, one end of a variable resistor is connected to +12 volts, and the other end is connected to ground. The middle terminal or wiper is connected to the positive terminal of a voltmeter, and the negative lead is connected to ground. If the wiper is moved to the upper end of the resistor, the voltmeter will indicate a potential of 12 volts. If the wiper is moved to the bottom, the voltmeter will indicate a value of 0 volts. The wiper can be adjusted to provide any value of voltage between 12 and 0 volts.

Schematic Symbols

Electrical schematics use symbols to represent the use of a resistor. Unfortunately, the symbol used to represent a resistor is not standard. Figure 1-25 illustrates several schematic symbols used to represent both fixed and variable resistors.

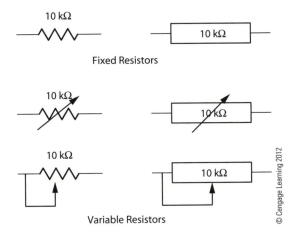

FIGURE 1-25 *Schematic symbols used to represent resistors*

UNIT 1 REVIEW QUESTIONS

1. The atoms of a material used as a conductor generally contain _____ valence electrons.

2. The atoms of a material used as an insulator generally contain _____ valence electrons.

3. The two most common materials used to produce semiconductor devices are _____ and _____.

4. What is a lattice structure? _____
_____
_____

5. How is a P-type material made? _____
_____
_____

6. How is an N-type material made? _____
_____
_____

7. Which type of semiconductor material can withstand the greatest amount of heat? _____
_____

8. All electronic components are formed from P- and N-type materials. What factors determine what kind of components will be formed? _____
_____
_____

9. Name three types of fixed resistors. _____
_____

10. What is the advantage of a metal film resistor over a carbon resistor? _____
_____

11. What is the advantage of a wire wound resistor? _____
_____

UNIT 1 REVIEW QUESTIONS

12. A half-watt 2000-ohm resistor has a current flow of 0.01 amps through it. Is this resistor operating within its power rating? _____

_____

13. A one watt 350-ohm resistor is connected to 24 volts. Is this resistor operating within its power rating? _____

_____

14. A resistor has color bands of orange, blue, yellow, gold. What is the resistance and tolerance of this resistor? _____

_____

15. A 10,000-ohm resistor has a tolerance of 5 percent. What are the minimum and maximum ratings of this resistor? _____

_____

16. Is 51,000 ohms a standard value for a 5 percent resistor? _____

_____

17. What is a potentiometer? _____

_____

UNIT 2

POWER RATING AND
HEAT SINKING COMPONENTS

OBJECTIVES

After studying this unit, the student should be able to:

- Discuss why heat sinks are necessary in electronic circuits.

- Discuss the use of thermal compound.

- Explain how thermal compound aids in the transfer of heat from the electronic component to the heat sink.

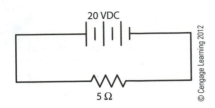

FIGURE 2-1 *Simple DC circuit*

$P = E \times I, P = 20 \times 4, P = 80 \text{ watts}$

$P = I^2 \times R, P = 4 \times 4 \times 5,$

$P = 80 \text{ watts}$

$P = \dfrac{E^2}{R}, P = \dfrac{20 \times 20}{5}, P = \dfrac{400}{5},$

$P = 80 \text{ watts}$

FIGURE 2-2 *Formulas used to compute the power in a circuit*

© Cengage Learning 2012

POWER RATING

Electrical and electronic components have a power rating measured in watts (w). This rating indicates the amount of heat that the component can dissipate before damage occurs. This can be found by using any one of three formulas, Power (P) = voltage (E) × current (I); Power = I^2 × resistance (R), Power = $\dfrac{E^2}{R}$. Consider the resistor shown in Figure 2-1.

In this example, a 5-ohm resistor is connected across 20 volts. A current of 4 amperes (A), will flow in this circuit.

$$\left(I = \frac{E}{R}, \quad I = \frac{20}{5}, \quad I = 4 \right).$$

To find the amount of heat this resistor must dissipate, use any of the three formulas, Figure 2-2.

When electrical components are connected into a circuit, care should be taken not to exceed their power ratings. If there is any doubt that the component can handle the heat it is expected to dissipate, the power should be computed by using one of the formulas shown in Figure 2-2. Forcing a component to dissipate more heat than it was designed for can only shorten its life.

HEAT SINKS

Some electronic components must be **heat sinked** in order to operate at their listed power ratings. Assume a transistor has a power rating of 30 watts. This power rating can be obtained only if the transistor is mounted on a proper heat sink and thermal compound used.

Heat sinks vary in size and shape, but have only one purpose. That purpose is to increase the surface area of the device connected to it. This permits air to wipe a greater area and remove heat at a faster rate.

When a component is mounted to a heat sink, **thermal compound** is generally used to ensure a good thermal contact between the device and the heat sink. There are two types of thermal compound in common use. One type is a greasy substance that is usually white in color. This compound is an excellent conductor of heat but is made from beryllium oxide, which is a deadly poison. When using this compound, care should be taken not to get it into the mouth or eyes. The second type in general use is a silicon grease that looks like a clear jelly. This compound is a good heat conductor and is less messy.

The surface of the device and heat sink may look perfectly flat to the naked eye. A closer examination, however, reveals that neither surface is flat. The surface appears similar to the surface shown in Figure 2-3.

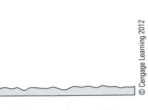

FIGURE 2-3 *Uneven surface of an electronic component*

If these two surfaces are joined together, they may make actual metal-to-metal contact at relatively few points, Figure 2-4. That is why thermal compound is used to fill in the gaps between the two surfaces and provide good thermal contact, Figure 2-5. The importance of using thermal compound when heat sinking components cannot be overstressed.

Some heat sinks are rather simple in design and are intended to be finger pressed onto a component, Figure 2-6. This type of heat sink does not generally require the application of thermal compound. Small heat sinks are generally used when a component is operated at its full power rating. A TO-3 transistor mounted on a heat sink is shown in Figure 2-7. Regardless of whether a component requires the use of a heat sink, be sure to keep the components operating within their power rating.

FIGURE 2-4 *Two components with uneven surfaces joined together*

FIGURE 2-5 *Heat sink compound used to fill the gaps*

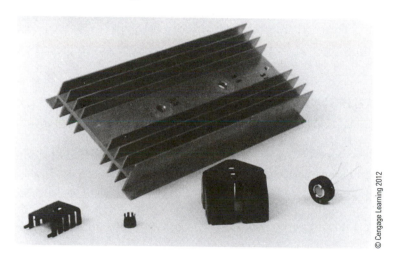

FIGURE 2-6 *Heat sinks*

FIGURE 2-7 *TO-3 transistor mounted on a heat sink*

UNIT 2 REVIEW QUESTIONS

1. An electronic component has a voltage drop of 30 volts when a current of 0.2 amp flows through it. How much power is being dissipated by this component? _____ watts

2. An electronic component has a voltage drop of 25 volts and an impedance of 10 ohms. How much power is being dissipated by this component? _____ watts

3. A component has a current of 0.6 amp flowing through it and a resistance of 100 ohms. How much power is being dissipated by this component? _____ watts

4. What function does a heat sink serve when it is attached to a component? _____
_____
_____

5. Why should thermal compound be used when attaching a component to a heat sink?
_____
_____

6. A 300-Ω resistor is to be connected to a 24-volt DC power source. Should the resistor be rated at ¼ watt, ½ watt, 1 watt, or 2 watts? _____

7. A transistor has a voltage drop of 12 volts across its collector-emitter circuit when a current of 250 ma (milliamperes) flows through it. How much power must this transistor dissipate? _____
_____

8. A 2.7-kΩ resistor has a current of 25 ma flowing through it. How much power must this resistor dissipate in the form of heat? _____
_____

9. A 1000-Ω resistor has a power rating of 10 watts. Can this resistor be connected to a 120-volt power source and operate within its power rating? _____
_____

10. A diode has a forward voltage drop of 0.7 volt. How much power must it dissipate in the form of heat when a current of 16 amps flows through it? _____
_____

UNIT 3

MEASURING INSTRUMENTS

OBJECTIVES

After studying this unit, the student should be able to:

- Discuss the operation of a digital multimeter.

- Measure voltages with a digital multimeter.

- Measure resistance with a digital multimeter.

- Determine the amount of current flow in a circuit.

- Discuss the operation of an oscilloscope.

- Discuss various oscilloscope controls.

- Connect an oscilloscope in a circuit.

- Interpret wave forms produced on the display of an oscilloscope.

The ability to measure values of voltage, current, and resistance in an electronic circuit is paramount for an industrial electrician. The two most common instruments to perform this job are the multimeter and the oscilloscope. Multimeters can be divided into two basic types: analog and digital. Analog meters are characterized by the fact that they contain a meter movement with a moving pointer, Figure 3-1. Although meters of this type have been used by electricians for many years, they are generally not the best choice for troubleshooting electronic circuits. Most analog voltmeters work on the principle of inserting resistance in series with the meter movement to affect a change in the full scale value of the meter, Figure 3-2. A common resistance value for a DC voltmeter is 20,000 ohms per volt. This means that if the meter is set for a full scale value of 15 volts, it will have an internal resistance of 300 kΩ (20,000 $\times$ 15). If the meter is set on the 15-volt range, it will require a current of 50 μA (microamperes) to operate the meter movement (15V/300 kΩ). If this meter is used to measure the voltage in a circuit that contains low resistance and a high current capacity, there is no problem. The amount of current needed to operate the meter movement is not sufficient to disrupt the circuit.

FIGURE 3-1 *Volt-ohm-milliampere meter with multirange selection*

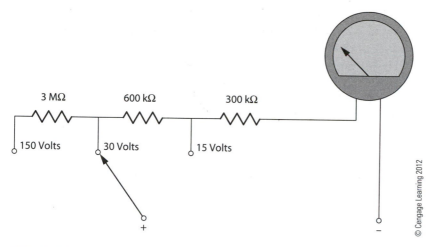

FIGURE 3-2 *Analog voltmeters change range by connecting resistance in series with the meter movement*

Now assume that this meter is to be used to measure the voltage drop across a 120-kΩ resistor in a circuit that contains a current flow of 100 μA. The meter should indicate a value of 12 volts (100 μA $\times$ 120 kΩ). When the meter is connected across the 120 kΩ resistor, a resistance of 300 kΩ is connected in parallel with a 120 kΩ resistor, Figure 3-3. The total resistance of this connection is 85,714.3 Ω.

$$R_t = \frac{R_1 \times R_2}{R_1 + R_2}$$

$$R_t = \frac{120,000 \times 300,000}{120,000 + 300,000}$$

$$R_t = \frac{36,000,000,000}{420,000}$$

$$R_t = 85,714.3 \ \Omega$$

The voltage drop across this connection is now 8.57 volts (85,714.3 Ω $\times$ 100 μA). Connecting the meter into the circuit has changed the resistance value to the point that it is not possible to measure the correct voltage drop across the 120-kΩ resistor. This is generally called "loading the circuit." As a general rule, analog multimeters do not contain enough internal resistance to accurately measure voltage values in an electronic circuit.

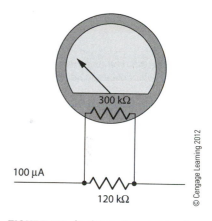

FIGURE 3-3 *Analog meters can load the circuit in a way such that the voltage measurement is not accurate.*

DIGITAL MULTIMETERS

Digital meters are characterized by the fact that they employ a display to indicate the value instead of a meter movement and pointer, Figure 3-4. The method of display is not the only difference

FIGURE 3-4 *Digital multimeter*

Courtesy of Advanced Test Products

between digital and analog meters. Most digital meters will exhibit the same amount of input resistance regardless of the range setting of the meter. This is accomplished by using a voltage divider circuit instead of inserting resistance in series with a meter movement, Figure 3-5. In the circuit shown, the voltmeter probes are connected across a voltage divider circuit with a total resistance of 10 megohms. In this way, the input resistance is the same regardless of the range setting of the meter. In the previous example, an analog meter with an internal resistance of 300 kΩ was connected across a 120-kΩ resistor. This connection caused extreme inaccuracy when measuring the voltage drop across the resistor. When the digital meter is connected across the 120-kΩ resistor, the total resistance becomes 118,577 Ω. The voltage drop is now 11.86 volts instead of 8.57 volts. The digital meter caused much less change to the circuit than the analog meter, permitting a more accurate voltage measurement to be taken.

Many digital multimeters are *autoranging*. This means that the meter will automatically determine the proper range setting when measuring voltage, current, or resistance. It is only necessary to set the meter to the desired quantity to be measured, such as DC volts, AC volts, ohms, DC mA, or AC mA. When using digital meters with the autorange feature, you must pay attention to the notations shown on the meter display. The meter will typically indicate a value plus a notation showing the proper engineering unit. The display shown in Figure 3-6 indicates a value of 365 millivolts. The notation "mV" indicates millivolts.

Digital Ohmmeters

Digital ohmmeters operate differently than analog ohmmeters also. Analog ohmmeters operate by adjusting the amount of current flow to the full scale value of the meter, Figure 3-7. To accomplish this, the meter probes are shorted together and a variable resistor is used to adjust the current. This is referred to as "zeroing the meter." The meter probes are then connected across the unknown resistance, Figure 3-8. The reduction of current causes the meter to indicate a value less than the full scale value. The amount of reduction in current is proportional to the amount of resistance inserted into the circuit.

Digital type ohmmeters employ a current generator to produce a known amount of current through the unknown resistance, Figure 3-9. A digital voltmeter is used to measure the voltage drop across the resistor. The voltage drop is proportional to the resistance value and the amount of current. Autoranging meters automatically adjust the current value to obtain an amount of current that will produce a voltage drop that is within the range of the voltmeter. As with autoranging voltmeters, you must be aware of the notation shown on

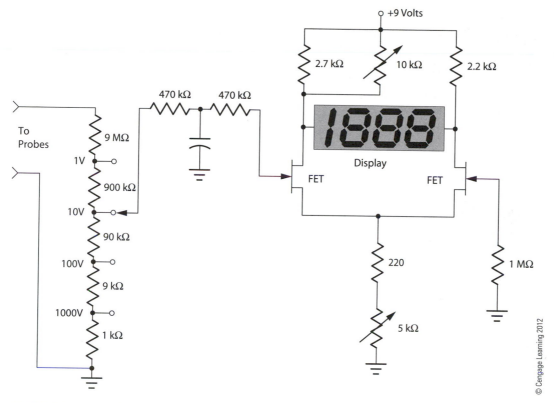

FIGURE 3-5 *Basic digital voltmeter circuit*

FIGURE 3-6 *The display indicates a value of 365 millivolts.*

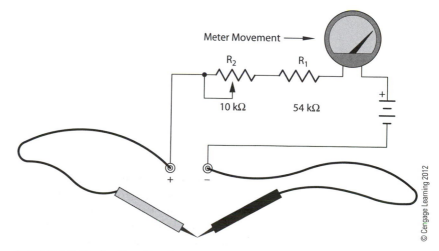

FIGURE 3-7 *Zeroing the ohmmeter*

FIGURE 3-8 *Measuring the resistance value*

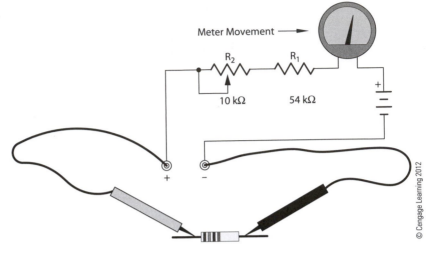

FIGURE 3-9 *Digital ohmmeters operate by measuring the voltage drop across a resistor when a known amount of current flows through it.*

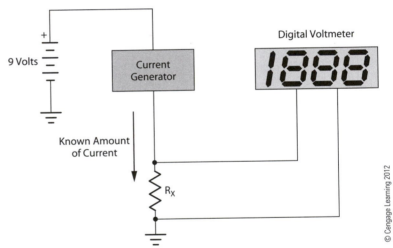

the meter display. Autoranging ohmmeters commonly display values in ohms (Ω), kilohms (kΩ), and megohms (MΩ).

Measuring Current

Most digital multimeters contain settings for both DC and AC amperes. When the current measuring function of the meter is to be employed, the meter must be connected in series with a load, just as any other ammeter, Figure 3-10.

Due to the nature of many electronic circuits, it is often not possible or convenient to break the circuit and connect a meter in series with a load. As a general rule, current in an electronic circuit is determined by measuring the voltage drop across a known resistance

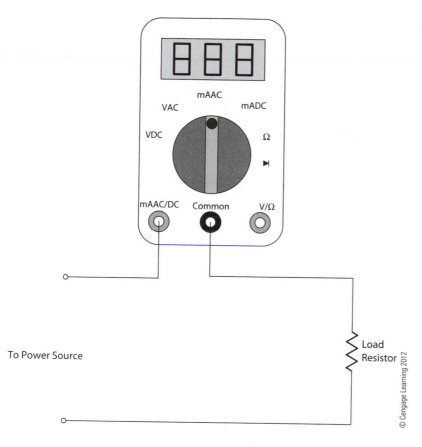

© Cengage Learning 2012

FIGURE 3-10 *The meter is connected in series with a load.*

value and then using Ohm's Law to determine the current. Assume, for example, that a resistor has a marked value of 43 kΩ. Now assume that the voltmeter indicates a voltage drop of 16.25 volts across the resistor. The current is 0.378 mA.

$$I = \frac{E}{R}$$

$$I = \frac{16.25}{43,000}$$

$$I = 0.0003779 \text{ or } 0.378 \text{ mA}$$

Semiconductor Test

Many digital multimeters have a special range that is used to test semiconductor devices. This range is indicated by a diode symbol (▶|). Although this setting operates similarly to the ohmmeter, it does not measure resistance. The semiconductor test range is used to test semiconductor devices such as diodes, transistors, LEDs, and so on. In order to test a semiconductor device, the meter must provide enough output voltage to overcome the forward voltage drop of the device. Silicon diodes and transistors, for example, require a voltage of 0.6 to 0.7 volt. Light-emitting diodes (LEDs) generally require a forward voltage of 1.7 to 2 volts. If less voltage is applied to the device, it will not conduct in the forward direction and will appear as defective when it is not. Ohmmeters typically do not provide the amount of voltage necessary to overcome the forward voltage of many semiconductor devices.

THE OSCILLOSCOPE

For many years the industrial electrician's measuring tools have been the volt-ohm-milliammeter (VOM) and the clamp-on ammeter. These old standbys are still good tools and are the best way to troubleshoot many of the circuits the industrial electrician encounters. However, many of the electronic control systems in today's industry produce voltage pulses that are meaningless to a VOM. In many instances, it is necessary not only to know the amount of voltage present at a particular point in a circuit, but also the length or duration of the pulse and its frequency. Some pulses may be less than 1 volt and last for only a millisecond. A VOM does not measure many of these things. Therefore, the oscilloscope must be used to learn what is actually happening in a circuit.

OSCILLOSCOPE BASICS

This unit is designed to teach some of the fundamentals of using an oscilloscope. It is not intended to make you an expert in its use. The first point to understand about the oscilloscope is that it is a voltmeter; it measures voltage. It does not measure current, resistance, or watts. The oscilloscope not only measures a voltage during a particular period of time, it creates a two-dimensional image, or picture, on its screen.

Some Important Parts of the Oscilloscope

The oscilloscope is divided into two main sections: the voltage section and the time base. The display of the oscilloscope is divided by vertical and horizontal lines, Figure 3-11. Voltage is measured on the vertical, or Y-, axis of the display and time is measured on the horizontal, or X-, axis of the display.

When using the VOM, a range switch permits the selection of a different range of voltages that will deflect the meter full scale—600 volts, 300 volts, or 60 volts, for instance. Changing voltage ranges permits much more accurate measurements of voltage. Trying to measure 24 volts on the 600-volt range will not move the meter enough to make any kind of accurate measurement. By changing to a range of 60 volts full scale, however, 24 volts can be read very accurately.

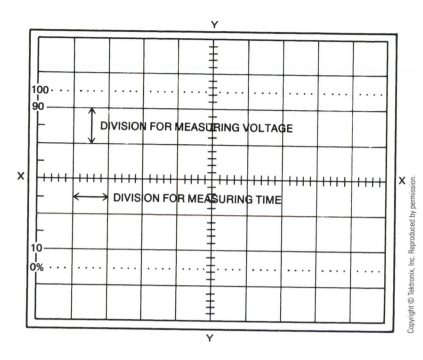

FIGURE 3-11 *Display of oscilloscope.*

The oscilloscope also has a voltage range switch, Figure 3-12. The voltage range switch on an oscilloscope selects volts per division instead of volts full scale. For instance, the voltage range switch shown in Figure 3-12 is set for 10 millivolts (mV) at the 1X position. This means that each of the lines in the vertical direction (the Y-axis) of the display, Figure 3-11, has a value of 10 millivolts per division. Assume the oscilloscope has been adjusted to 0 volts on the centerline of the display. The probe is then removed from ground and connected to the circuit to be tested. If the trace rises above the centerline, the voltage is positive with respect to ground. If the trace drops below the centerline, the voltage is negative with respect to ground. If the oscilloscope probe is connected to a positive voltage of 30 millivolts, the trace rises to the position marked (A) in Figure 3-13.

If the probe is connected to a negative 30 millivolts, the trace falls to the position marked (B) in Figure 3-13. Notice that the oscilloscope can display a negative voltage as easily as it can display a positive voltage. If the range is changed to 20 volts per division, (A) in Figure 3-13 will display a value of 60 volts positive.

The next part of the oscilloscope to become familiar with is the time base, Figure 3-14. The time base is calibrated in seconds per division and has range values from seconds to microseconds. The time base controls the value of the divisions of the lines in the horizontal direction. For instance, if the time base is set for 5 milliseconds per

FIGURE 3-12 *Voltage range control*

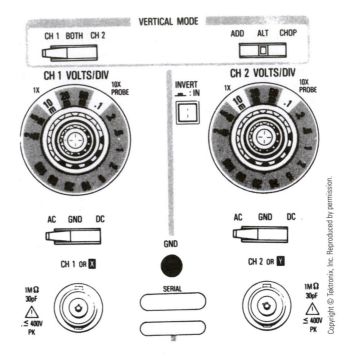

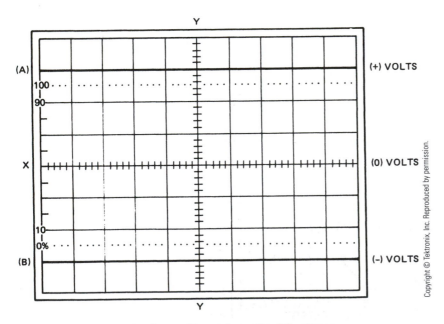

FIGURE 3-13 *Display showing positive and negative DC voltages*

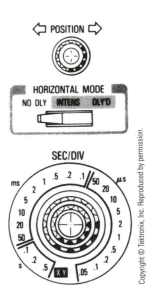

FIGURE 3-14 *Time base*

division, the trace will sweep from one division to the next in 5 milliseconds. With the time base set in this position, it will take the trace 50 milliseconds (ms) to sweep from one side of the display screen to the other. If the time base is set for 2 microseconds per division, the trace will sweep the screen in 20 microseconds.

The oscilloscope not only measures the value of voltage, it also measures the voltage with respect to time. Because the oscilloscope can measure the time to complete one cycle of AC voltage, its frequency can be computed. This is done by dividing 1 by the time it takes to complete one cycle, $F = 1/t$. Assume that the time base is set for 0.5 milliseconds per division and the voltage range is set for 20 volts per division. If the oscilloscope has been set so that the centerline of the display is 0 volts, the AC waveform shown in Figure 3-15 will have a peak value of 60 volts. Notice that the oscilloscope displays the peak value of voltage and not the root-mean-square (rms) or effective value, which is measured by an AC voltmeter. To measure the frequency, count the time it took to complete one full cycle. The waveforms shown in Figure 3-15 took 4 ms. Therefore, the frequency is $1/0.004$ s $= 250$ (hertz) Hz.

Most oscilloscopes use a probe that acts as an attenuator. An **attenuator** is a device that divides or makes smaller the input signal. An attenuated probe is used to permit higher voltage readings than are normally possible. Most attenuated probes are 10:1. This means

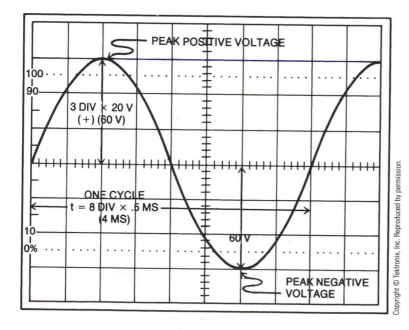

FIGURE 3-15 *Display showing AC sine wave.*

that if the voltage range switch is set for 5 volts per division, the display is read as 50 volts per division. If the voltage range switch is set for 2 volts per division, each division on the display has a value of 20 volts per division.

Probe attenuators are made in various styles by manufacturers. Some probes have the attenuator located in the head, whereas others have it located at the scope input. Regardless of the type of attenuator used, it may have to be compensated or adjusted. In fact, probe compensation should be checked frequently. Manufacturers also use different methods for compensating (adjusting) their probes, so follow the procedures given in the operator's manual for the oscilloscope being used.

Some Common Controls

Become familiar with the more common controls found on an oscilloscope, Figure 3-16.

- **Power:** The power switch is used to turn the oscilloscope on or off (1).

- **Beam finder:** This control is used to locate the position of the trace if it is off the display. The beam-finder button indicates the approximate location of the trace and will help in moving the position controls to get the trace back on the screen (2).

FIGURE 3-16 *Front of Tektronix Model 2213 oscilloscope.*

- **Probe adjust:** This is a reference voltage point used to compensate the probe. Most probe adjust points produce a square-wave signal at about 0.5 volt (3).

- **Intensity and focus:** The intensity control adjusts the brightness of the trace.

  The focus control sharpens the image of the trace (4).

- **Vertical position:** Adjust the trace up or down on the display. If a dual trace scope (an oscilloscope that can display two separate voltages at the same time) is being used, there are two vertical position controls (5).

- **Ch 1–Both–Ch 2:** This control permits the selection of either channel 1, channel 2, or both at the same time (6).

- **Add–Alt–Chop:** This control is active only when both traces are being displayed at the same time. The add mode adds the two waves together. The alt mode alternates the sweeps between channel 1 and channel 2. The chop mode alternates several times during one sweep. This makes the display appear to be more stable. The chop mode is used more frequently when observing two traces at the same time (7).

- **AC–GND–DC:** The alternating current (AC) mode is used to block a direct current (DC) voltage when only the AC part of the voltage is to be seen. Assume an AC voltage of a few millivolts is riding on a DC voltage of a hundred volts. If the voltage range is set high enough to permit 100 volts of direct current to be seen on the display, the AC voltage will not be visible. The AC section of this switch inserts a capacitor in series with the probe. The capacitor blocks the DC voltage and permits the AC voltage to pass. Because the 100 DC volts have been blocked, the voltage range can be adjusted for millivolts per division and the small AC signal can be seen. The GND (ground) section of the switch grounds the input so the sweep can be adjusted for 0 volts at any position on the display. The ground switch provides ground at the scope but does not ground the probe. This means that the ground switch can be connected to a live circuit without a problem. The DC section permits observation of the voltage to which the probe is connected (8).

- **Horizontal position:** This control adjusts the position of the trace from left to right (9).

- **Auto–normal:** The auto–normal control determines whether the time base will be triggered automatically or if it is to be operated in a free-running mode. If it is operated in the normal setting, the trigger signal is taken from the line to which the

CAUTION!

© iStock 2012

Be careful not to leave a bright spot on the display, because it will burn a spot on the face of the cathode-ray tube (CRT). This burned spot results in permanent damage to the CRT.

probe is connected. The scope is generally operated with the trigger set in the automatic position (10).

- **Level:** This control determines the amplitude that the signal must be before the scope triggers (11).

- **Slope:** The slope permits the scope to trigger on the positive or negative half of the waveform (12).

- **Int–Line–Ext:** The scope is generally operated in the Int (internal) mode. In this mode, the trigger signal is provided by the scope. In the line mode, the trigger signal is provided from a sample of the line. The Ext (external) mode permits an external trigger signal to be applied (13).

These are not all the controls shown on the oscilloscope in Figure 3-16, but they are the major ones. Most oscilloscopes contain these controls.

INTERPRETING WAVEFORMS

After learning to operate the controls on the scope, interpreting the waveforms shown on the display is learned. It generally takes experience and practice to become proficient in the use of the oscilloscope. When using the oscilloscope, remember that the display is the product of voltage and time. A camera can be used to photograph the various waveforms. These photographs can then be used for comparison of the waveforms as seen on the oscilloscope. Many oscilloscopes permit an image of the waveform to be downloaded into a computer so that it can be stored or analyzed.

In Figure 3-17, assume that the voltage range has been set for 0.5 volt per division, the time base is set at 2 milliseconds per division, and 0 volts has been set on the centerline of the display. The waveform shown is a square wave. The display shows that the voltage rises in the positive direction to a value of 1.4 volts and remains there for 2 ms. It then drops to 1.4 volts negative and remains there for 2 ms before going back to positive. Because the voltage changes between positive and negative, it is an AC voltage. The length of one cycle is 4 ms, so the frequency is 1/0.004 = 250 Hz.

In Figure 3-18, the oscilloscope has been set for 50 millivolts (mV) per division and 20 microseconds (μs) per division. The display shows a voltage that is negative to the ground lead of the probe, and has a peak value of 150 millivolts (mV). The pulse waveform crosses zero every 20 microseconds, therefore the pulse rate is 1/0.000020 = 50 kHz (kilohertz). The voltage shown is DC because it never crosses the zero reference and goes in the positive direction.

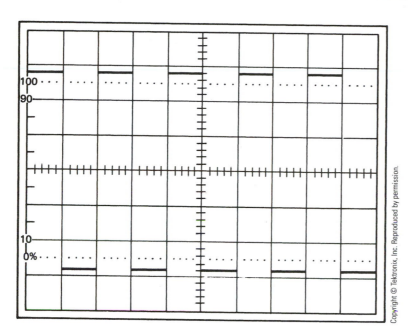

FIGURE 3-17 *Display showing AC square wave.*

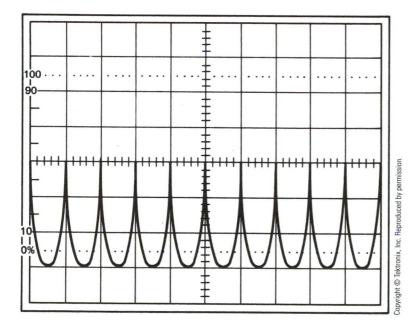

FIGURE 3-18 *Display showing DC voltage in the negative direction.*

In Figure 3-19, assume that the scope is set at 50 volts per division and 0.1 ms per division. The pulse waveform rises to a value of 150 volts in the positive direction and then drops to about 25 volts. It remains at 25 volts for 0.15 milliseconds before dropping back to 0 volts. It remains at 0 volts for 0.3 milliseconds before the waveform

FIGURE 3-19 *Display showing a DC waveform.*

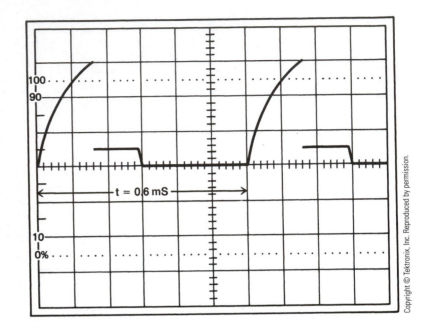

starts over again. The voltage shown is DC because it remains in the positive direction. To find the pulse rate, measure from the beginning of one pulse to the beginning of the next pulse. This is the period of one complete pulse. The length of one pulse period is 0.6 milliseconds. Therefore, the pulse rate is 1/0.0006 = 1666 Hz.

Learning to interpret the waveforms seen on the display takes time and practice. It is worth the effort, however, because it is the only way to understand what is happening in many electronic circuits.

OSCILLOSCOPE PRECAUTIONS

When using an oscilloscope, certain precautions must be taken when connecting the probe to a live circuit. Most oscilloscopes are powered from a 120-volt AC source. Connection to the power line is generally made with a standard three-prong plug, as shown in Figure 3-20. The black conductor of the chord is connected to the underground side of the circuit, and the white conductor is connected to ground and is the neutral conductor. The green conductor is connected to the metal case of the oscilloscope and is used as the safety grounding conductor.

The probe end of the oscilloscope contains two conductors. One is connected to the input and in most cases has an impedance of a million ohms or more. The other conductor, however, is connected directly to ground. If the oscilloscope should be used to test a circuit that has one side grounded and the other side ungrounded, care must

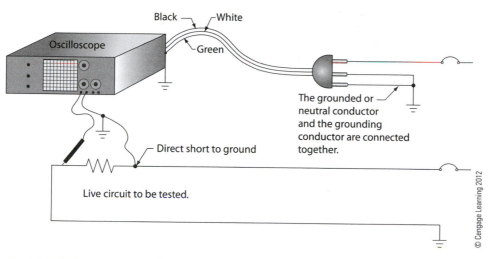

Black ──┐   ┌─White
        └─Green

Oscilloscope

The grounded or —
neutral conductor
and the grounding
conductor are connected
together.

─ Direct short to ground

Live circuit to be tested.

© Cengage Learning 2012

FIGURE 3-20 *The grounded probe of the oscilloscope is connected to the ungrounded circuit conductor.*

be taken to ensure that the grounded conductor of the probe is connected to the grounded side of the circuit. If the grounded conductor of the probe should be connected to the ungrounded side of the circuit, a direct short to ground through the probe lead and case of the oscilloscope will exist.

UNIT 3 REVIEW QUESTIONS

1. What electrical quantity such as voltage, current, resistance, or watts is the oscilloscope used to measure? _____

2. What quantity is measured on the X-axis of an oscilloscope? _____

_____

3. What quantity is measured on the Y-axis of an oscilloscope? _____

_____

4. An AC waveform completes one cycle in 200 microseconds. What is the frequency of the voltage?

_____

5. Refer to Figure 3-21. An oscilloscope probe is connected to a 10:1 attenuator. The volts-per-division switch is set for a value of 5 volts. What is the peak-to-peak voltage shown on the display? _____

_____

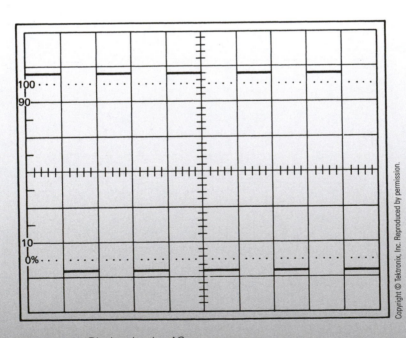

FIGURE 3-21 *Display showing AC square wave.*

UNIT 3 REVIEW QUESTIONS

6. Refer to Figure 3-22. An oscilloscope probe is connected to a 100:1 attenuator. The volts-per-division switch is set at 0.1. The time base is set for 20 μs per division. What is the peak voltage and the frequency of the waveform shown? _____

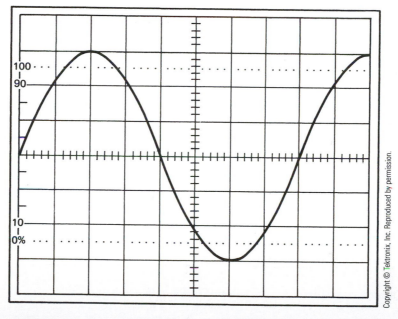

FIGURE 3-22 *Display showing AC sine wave.*

7. What is the beam finder used for? _____
_____

8. What is the difference between the alternate and chop mode? _____
_____
_____
_____

9. Why should a bright spot not be left on the display of the oscilloscope? _____
_____

UNIT 3 REVIEW QUESTIONS

10. What is the function of the slope control? _____
_____
_____

11. An analog voltmeter has a resistance of 20,000 ohms per volts. What is the total resistance of the meter circuit when the meter is set on the 250-volt range? _____
_____

12. Assume that the meter in question 11 is set on the 12-volt range. Now assume that the meter is connected across a resistor with a value of 10 kΩ. What is the total resistance of the circuit with the meter connected across the resistor? _____
_____

13. What type of meter measures the voltage drop across a resistor to determine its resistance value?
_____
_____

14. A digital voltmeter indicates a voltage drop of 4.6 volts when connected across a 2.2-kΩ resistor. How much current is flowing through the resistor? _____
_____

15. What type of meter exhibits the same amount of input resistance regardless of the range setting?
_____
_____

UNIT 4

JUNCTION DIODES

OBJECTIVES

After studying this unit, the student should be able to:

- Discuss the operation of a junction diode.

- Explain forward and reverse bias.

- Draw the schematic symbol for a junction diode.

- Test a junction diode with an ohmmeter.

- Explain the differences between the conventional current flow theory and the electron flow theory.

Junction diodes are often referred to as pn junctions because they are made by joining a piece of P-type and a piece of N-type semiconductor material together, Figure 4-1. Basically, diodes are electrical check valves. They permit current to flow through them in one direction only.

To understand the operation of a junction diode, it will be necessary to return to the study of semiconductor materials. Recall from Unit 1 that a P-type material has an excess of holes in its structure and an N-type material has an excess of electrons. When P- and N-type materials are joined, a barrier charge is formed at the junction that prevents electrons from crossing into the P region and holes from crossing into the n region. Now assume that a battery and a current-limiting resistor are connected to the leads of a PN junction. Also assume that the negative terminal of the battery is connected to the P-type material and the positive battery terminal is connected to the N-type material, Figure 4-2. The majority of current carriers in P-type material are holes, and the majority of current carriers in an N-type material are electrons. Positive holes will be attracted to the negative polarity and negative electrons will be attracted to the positive polarity. In this condition, the PN junction is reverse biased and current will not flow through it.

If the battery voltage is reversed, and the battery has a voltage greater than the barrier potential (0.7 volt for silicon and 0.4 volt for germanium), electrons will be attracted to the positive polarity and drift across the junction into the P region. Holes, attracted to the negative battery polarity, will drift into the N region, Figure 4-3. When an electron reaches the positive terminal, it leaves the P region and flows to the battery. When the electron leaves the P region, a hole is formed. When a hole reaches the negative terminal, it is replaced with an electron from the battery. There are always the same

FIGURE 4-1 *A PN junction is formed by joining a piece of P-type and a piece of N-type semiconductor material together.*

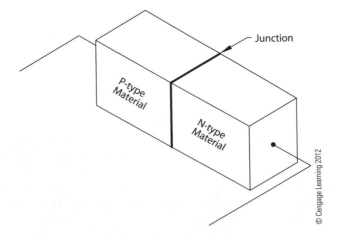

Junction

P-type Material

N-type Material

© Cengage Learning 2012

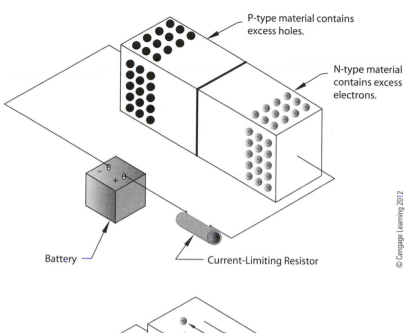

P-type material contains excess holes.

N-type material contains excess electrons.

Battery

Current-Limiting Resistor

© Cengage Learning 2012

FIGURE 4-2 *The PN junction will not conduct when reverse biased.*

FIGURE 4-3 *Current flows through the PN junction when it is forward biased.*

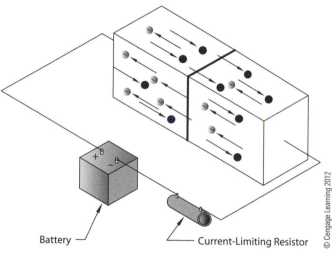

Battery

Current-Limiting Resistor

© Cengage Learning 2012

number of holes and electrons in the P and N regions. When the battery is connected in this manner, the diode is forward biased and current will flow through it.

The diode symbol is shown in Figure 4-4. This symbol was actually developed to illustrate the current path through a vacuum tube. The anode or plate is connected to the more positive voltage, and the cathode is connected to the more negative voltage. The symbol depicts electrons leaving the cathode and being attracted to the anode or plate, Figure 4-5. In order for the diode to conduct, the anode must be connected to the most positive voltage, and the cathode to the most negative, Figure 4-6.

FIGURE 4-4 *Diode symbol*

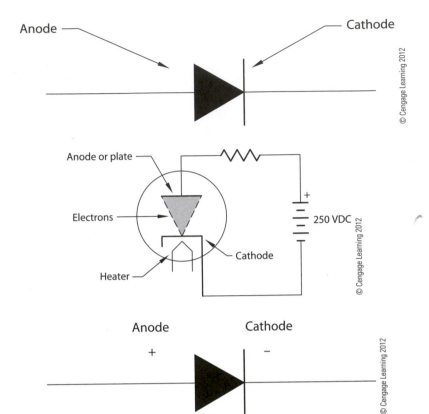

Anode ——⟋   ⟍—— Cathode

© Cengage Learning 2012

FIGURE 4-5 *Diode vacuum tube*

Anode or plate ——⟋

Electrons ——

Heater ——   Cathode

250 VDC

© Cengage Learning 2012

FIGURE 4-6 *The diode must be connected to the proper polarity before it will conduct current.*

Anode   Cathode

+   −

© Cengage Learning 2012

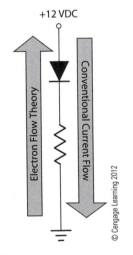

+12 VDC

Electron Flow Theory

Conventional Current Flow

© Cengage Learning 2012

FIGURE 4-7 *The arrow points in the direction of conventional current flow.*

CONVENTIONAL CURRENT FLOW

Many people prefer to use the conventional current flow theory when determining the current path through an electronic circuit. The conventional current flow theory assumes that current flows from positive to negative instead of negative to positive. There are several reasons for using this theory. Most electronic circuits use the negative power source as ground. The positive power source is considered to be above ground. Most people tend to think of something flowing down rather than up. Electronic diagrams are drawn to assume that current flows from positive to negative. Another reason that many people prefer the conventional current flow theory is that the arrow symbols of the components point in the direction of conventional current flow.

In the circuit shown in Figure 4-7, a diode is connected to ground through a current-limiting resistor. The diode in this example is forward biased and current will flow through it. If the electron flow theory is used to trace the current path, it must be assumed that electrons leave the ground terminal and flow through the diode against the arrow to the positive terminal. If the conventional current flow theory is used, it is assumed that current leaves the positive terminal and flows in the direction of the arrow to the ground terminal. In reality, it really does not make much difference which theory is used, as long as you are consistent. Because it is common practice to use both of these theories in explaining circuit operations, both will be used in this text. Some manufacturers of electronic equipment offer explanations using the conventional current flow theory, and others give explanations using the electron flow theory. Any student of electronics should become comfortable using either theory to analyze a circuit.

RECTIFIER DIODES

Common junction diodes are often referred to as **rectifier** diodes. They are used in many applications, such as the construction of recti-fiers, which convert alternating current into direct current. They can be obtained in a variety of case sizes and styles. They have ratings that range from hundreds of amperes to milliamperes and voltage ratings from low voltage to several thousand volts. Some rectifier-type di-odes can operate at only low frequencies, whereas others can operate in the microwave range. Diodes that can operate at high frequencies are often referred to as signal diodes. Their electrical characteristics are generally:

V_R—Volts reverse (expressed as PIV, peak inverse voltage, or PRV, peak reverse voltage). This rating indicates the amount of voltage the diode can block before conducting in the reverse direction.

V_F—Volts forward. Expresses the average voltage drop in the forward direction. Most rectifier diodes will have a forward voltage drop of about 1 volt.

I_F—Average forward current the diode can conduct.

I_R—Average current in the reverse direction the diode can withstand without being damaged. For rectifier-type diodes, this value is generally very low, typically in the micro- or nano-ampere range.

t_{rr}—Reverse recovery time. The amount of time in seconds required for the diode to turn off. The t_{rr} rating determines the operating frequency of the diode. This rating can be extremely important, depending on the type of circuit application.

I_{FS}—Forward surge current. The amount of short duration cur-rent the diode can withstand. Diodes often have to withstand large amounts of current flow when power is first applied. A typical example of this is if a diode is connected directly to a ca-pacitor. When the switch in Figure 4-8 is closed, a large amount

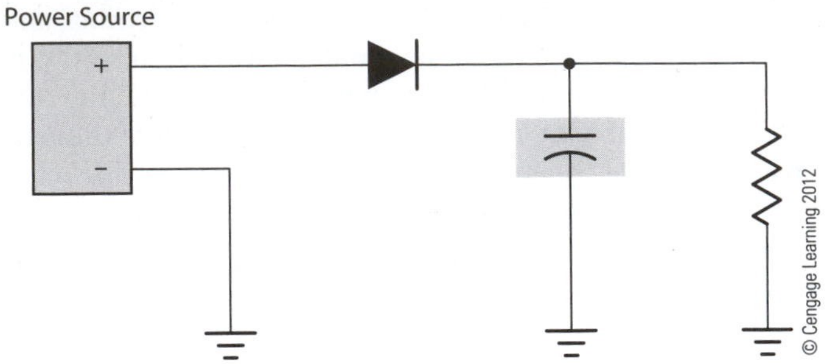

FIGURE 4-8 *A large amount of current will flow until the capacitor is charged.*

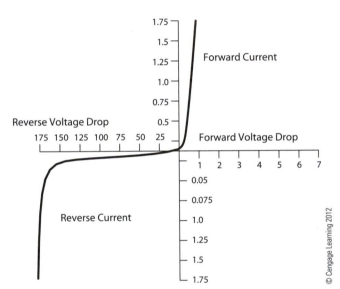

FIGURE 4-9 *Characteristic curve of a typical rectifier diode*

of current will flow until the capacitor is charged. As far as the power source is concerned, the capacitor is a short to ground until it becomes charged. The initial surge current in this circuit is limited only by the impedance of the circuit (internal impedance of the power supply, resistance of the wires, etc.).

P_D—The maximum power dissipation expressed in watts or milliwatts.

Dynamic Impedance

The forward voltage drop of most types of diodes remains relatively constant regardless of the amount of current flow. The forward voltage drop of most rectifier diodes is approximately 1 volt, Figure 4-9. In order for the diode to maintain a relatively constant voltage drop as current flow increases, the internal impedance of the device must change. For example, according to the characteristic curve shown in Figure 4-9, the diode has a forward voltage drop of approximately

0.5 volt when a current of 0.5 ampere is flowing through it. Its internal impedance at this point is 1 ohm (0.5/0.5). At a current of 1.75 amperes, the diode has a voltage drop of approximately 1 volt. The internal impedance at this point is 0.57 ohm (1/1.75).

Reverse Breakdown Current

Rectifier diodes are generally considered to be an open circuit when reverse biased. In reality, some amount of leakage current will flow in the reverse direction. This leakage current is typically in the microampere or nanoampere range. The leakage current will increase slightly with an increase of reverse voltage. If the PIV or PRV rating is exceeded, however, reverse current will suddenly become excessive. Rectifier diodes are generally destroyed when this happens because of the amount of heat generated. Assume that a diode has a PRV rating of 150 volts. Now assume that then the voltage rating is exceeded as a current of 1 ampere flows through the diode. The diode must dissipate 150 watts at this point (150 × 1).

Diode Markings

Diodes are generally marked in one of two ways. Diodes with a relatively large case size often have a diode symbol printed on the case to indicate which end is the anode and which is the cathode. Small size diodes have a line at one end of the device to indicate the cathode end, Figure 4-10. The line represents the line in front of the arrow.

The electrical characteristics of a diode can sometimes be determined by numbers on the diode. If the diode is a registered

FIGURE 4-10 *Different size diodes are marked in different ways.*

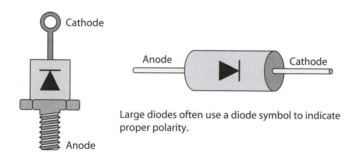

Large diodes often use a diode symbol to indicate proper polarity.

Small diodes use a line at one end to indicate the cathode. The cathode is the straight line in front of the arrow symbol.

© Cengage Learning 2012

FIGURE 4-11 *Electrical characteristics of diodes listed in the 1N registry*

Type	Material	Identification	PRV (Volts)	V_F	I_F	I_R	trr (s) m
1N315	G	R	300	0.48	0.075	0.3 ma	25
1N1341	S	R	50	1.6	6.0	4.0 ma	150
1N223A	S	R	800		1.0	0.003 ma	20
1N2316	S	R	350	1.1	35	20 ma	300
1N4004	S	R	400	1.1	1.0	0.03 ma	30
1N4374	S	R	1500	1.75	0.75	0.1 ma	15
1N4596	S	R	1400	1.35	150	3.5 ma	3000

© Cengage Learning 2012

device, it will contain a number that begins with 1N. A table showing the electrical characteristics of several diodes is shown in Figure 4-11. The material column indicates the device is made of silicon or germanium. The identification column indicates the specific type of diode, such as R for rectifier; DZ for diode, zener; and DS for diode, signal.

Some diodes contain "in-house" part numbers instead of 1N numbers. In-house numbers are numbers used by manufacturers. A manufacturer's reference book is required to determine the electrical characteristics. Motorola, for example, often uses numbers that begin with 1M, MR, MPZ, MRA, MZ, and MZC to identify in-house diodes. Hewlett Packard often uses numbers that begin with HSMP. In-house numbers do not mean that the components are of lower quality than registered devices; they are simply devices the manufacturers did not choose to register.

Equipment manufacturers often remove the existing numbers on components and replace them with their own numbers. As a general rule, the characteristics of these components cannot be determined by anyone other than the manufacturer. This forces the customer to purchase replacement parts from the manufacturer. A photograph showing diodes in different case styles is shown in Figure 4-12.

Testing Diodes

Most diodes can be tested with an ohmmeter. The exception to this is light-emitting diodes. Some ohmmeters do not provide enough voltage at the test leads to cause light-emitting diodes and some other types of electronic components to conduct in the forward direction. To overcome this problem, some ohmmeters contain a diode symbol on them.

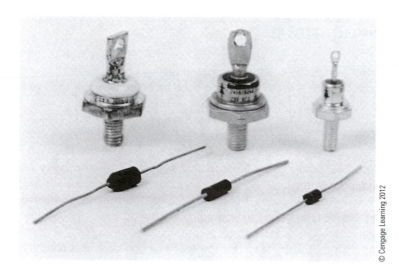

FIGURE 4-12 *Diodes shown in different case styles*

© Cengage Learning 2012

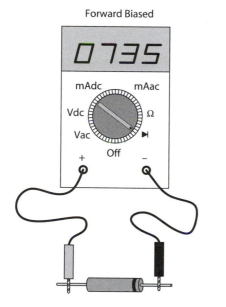
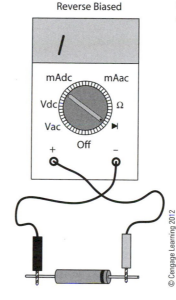

FIGURE 4-13 *Testing a diode with a digital multimeter*

© Cengage Learning 2012

When the selector switch is set to the diode symbol, the meter provides a high enough output voltage to turn on most solid-state devices.

Testing a common rectifier diode is generally accomplished by setting a digital meter on the semiconductor test range. The meter leads are then used to test the diode for continuity in both directions. The meter should indicate continuity through the diode in one direction only, Figure 4-13. If the meter does not have a semiconductor test range, the ohmmeter section can generally be employed to make this test. The forward voltage of rectifier diodes is low enough that a common ohmmeter can supply enough voltage to cause conduction.

APPLICATIONS

The diode has found many thousands of uses in the home and industry. Two circuits using a single diode will be discussed. The first circuit, Figure 4-14, is a dimmer control for a lamp. The switch is a single-pole double-throw with a center-off position. When the switch is in the center position, there is no power applied to the lamp. When the movable contact makes connection with the upper stationary contact, full voltage is connected to the lamp and it burns at full brightness. If the movable contact is changed to permit it to make connection with the lower stationary contact, the diode is connected in series with the lamp. Because the diode permits current to flow through it in only one direction, half of the AC waveform is blocked during each cycle. This permits only half the voltage to be applied to the lamp, causing it to burn at half brightness. The advantage of this type of dimmer control is that standard light bulbs can be used. This dimmer circuit does not require the use of the more expensive 3-way bulbs.

The second circuit to be discussed is shown in Figure 4-15. This circuit is very similar to the circuit shown in Figure 4-14, with the exception of a different type of switch and the addition of a current-limiting resistor. This circuit is used to provide a dynamic brake for a small AC induction motor. AC induction motors can be braked by applying direct current to their stator windings. A diode permits

FIGURE 4-14 *Light dimmer control*

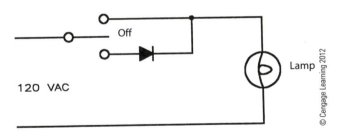

FIGURE 4-15 *Dynamic braking circuit for an AC induction motor*

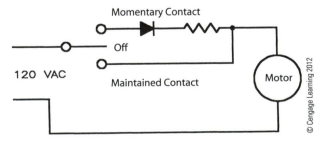

current to flow through it in only one direction. Because the current never reverses direction, it is direct current.

The switch used is a single-pole double-throw with a center-off position. When the switch is thrown in one direction, it maintains contact. When the switch is thrown in the other direction, it is a momentary contact. The circuit is so constructed that when the switch is thrown in the maintained contact position, the AC motor is connected to the power line. When the switch is thrown in the momentary contact position, it connects the diode and the current-limiting resistor in series with the AC induction motor. This applies direct current to the stator winding and brakes the motor. The resistor is sized to limit the amount of current flow to a safe value for both the diode and the winding of the AC motor.

UNIT 4 REVIEW QUESTIONS

1. How many elements of semiconductor material are contained in a diode? _____

_____

2. What are the two most common materials used in the production of diodes? _____

_____

3. If a diode is made from silicon, how many volts does it take to turn it on so that it will conduct
current in the forward direction? _____

_____

4. Must the voltage connected to the anode of a diode be positive or negative to make it forward
biased? _____

_____

5. What does the PIV rating on a diode mean? _____

_____

6. Explain how to test a diode with an ohmmeter. _____

_____

_____

UNIT 5

LIGHT-EMITTING DIODES (LEDs) AND PHOTODIODES

OBJECTIVES

After studying this unit, the student should be able to:

- Discuss the operation of a light-emitting diode (LED).

- Compute the resistance needed for connecting an LED into a circuit.

- Connect an LED in a circuit.

- Discuss the differences between light-emitting diodes and photodiodes.

- Draw the schematic symbols for LED and photodiodes.

LIGHT-EMITTING DIODES

Light-emitting diodes (LEDs) are among the most common devices found in the electrical/electronics field. They were first introduced in 1962. Early LEDs emitted low-intensity red light and were used as indicator lights for many years. Modern versions, however, are available across a wide spectrum of visible, ultraviolet, and infrared wavelengths with very high brightness. As their name implies, light-emitting diodes are basically diodes that perform a special function. When electrons in the N-type material are combined with holes in the P-type material, energy is released in the form of photons. The color of the light is determined by the energy level of the photons. The energy level of the photons is determined by the *energy gap* of the semiconductor material. The energy gap of the semiconductor is determined by the material used.

The basic light-emitting diode is formed by joining gallium arsenide (GaAs) or gallium phosphide (GaP). These two solutions can be combined to form a solid solution called gallium arsenide phosphide (GaAsP). Different colors can be produced by adding other compounds. The chart in Figure 5-1 shows different colored LEDs, the wavelength of light in nanometers, the forward voltage drop, and the materials used to construct the diode.

One of the great advancements in LED technology was the development of white LEDs. There are two primary ways of producing white LEDs. One employs the use of three individual LEDs that emit the primary colors of red, green, and blue (RGB). The most common method employs a blue and ultraviolet LED with a yellow phosphor coating, much like the coating on fluorescent lights. The phosphor coating changes the monochromatic light produced by the

FIGURE 5-1 *The color on an LED is determined by the material it is made from.*

Color	Material	Dopant	Wavelength (nm)
IR	GaAs	Si	900–1020
IR	GaAs	Zn	900
Red	GaP	Zn, O	700
Red	GaAsP	—	650
Orange	GaAsP	N	632
Yellow	GaP	N, N	590
Yellow	GaAsP	N	589
Green	GaP	N	570
Blue	SiC	—	490
Blue	ZnSe	—	490

blue/UV LED into a broad-spectrum white light. It is possible to produce different spectrums that produce different shades of white. White LEDs are very popular for use in portable lights because of their low power consumption. They are available in sizes up to 3 watts to produce bright portable lights.

LED Characteristics

The electrical characteristics of LEDs vary considerably from those of the common junction or rectifier diode. Junction diodes have a forward voltage drop of about 0.7 volt for silicon and 0.4 volt for germanium. LEDs have a forward voltage drop of about 1.7 volts or greater, depending on the material the diode is made of. Most LEDs operate at about 20 mA or less current. A chart showing the typical forward voltage drop of different diodes is shown in Figure 5-2. Junction diodes typically have a PIV rating of 100 volts or greater,

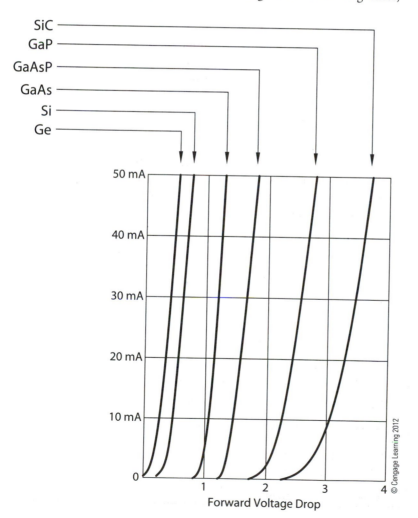

FIGURE 5-2 *Forward voltage and current characteristics of diodes*

but LEDs have a typical PIV rating of about 5 volts. For this reason, when light-emitting diodes are used in applications where they are intended to block any amount of reverse voltage, they are connected in series with a junction diode.

Testing LEDs

Light-emitting diodes can be tested in a manner similar to that of testing a junction diode. The LED is a rectifier and should permit current to flow through it in one direction only. When testing an LED with an ohmmeter, the meter must be capable of supplying enough voltage to overcome the forward conduction voltage of about 1.7 volts or higher. The meter, however, must not supply a voltage that is higher than the reverse breakdown voltage. The schematic symbol for a light-emitting diode is shown in Figure 5-3. Some symbols use a straight arrow as shown in Figure 5-3, and others use a lightning arrow as shown in Figure 5-4. The lightning arrow symbol is employed to help prevent the arrow from being confused with a lead attached to the device. The important part of the symbol is that the arrow is pointing away from the diode. This indicates that light is being emitted, or given off, by the diode.

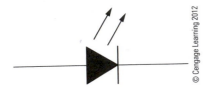

FIGURE 5-3 *Schematic symbol for a light-emitting diode*

© Cengage Learning 2012

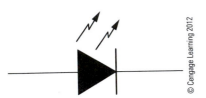

FIGURE 5-4 *LED symbol using lightning arrows*

© Cengage Learning 2012

LED Lead Identification

LEDs are housed in many different case styles. Regardless of the case style, however, there is generally some method of identifying which lead is the cathode and which is the anode. The case of most LEDs will have a flat side that is located closer to the cathode lead, Figure 5-5. Also, the cathode lead is generally shorter.

Seven-Segment Displays

A very common device that employs the use of LEDs is the seven-segment display, Figure 5-6. The display actually contains eight LEDs, each segment plus the decimal point. Common cathode displays have all the cathodes connected together to form a common point. The display is energized by connecting a more positive voltage to the anode lead or each segment. Common anode displays are energized by connecting the appropriate cathode lead to a more negative voltage (generally ground). The seven-segment display can be used to display any number from 0 to 9.

Connecting the LED in a Circuit

When used in a circuit, the LED generally operates with a current of about 20 mA (0.020 A) or less. Assume that an LED is to be connected in a 12-VDC circuit and is to have a current draw of approximately

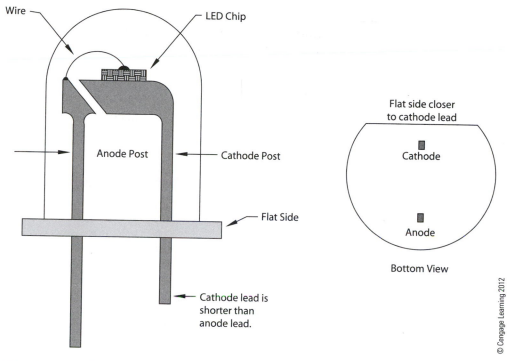

FIGURE 5-5 *Identifying the leads of an LED*

20 mA. This LED must have a current-limiting resistor connected in series with it. Ohm's Law can be used to determine what size resistor should be connected in the circuit.

$$R = \frac{E}{I}$$

$$R = \frac{12}{0.020}$$

$$R = 600\,\Omega$$

The nearest standard size resistor without going below $600\,\Omega$ is $620\,\Omega$. A $620\,\Omega$ resistor would be connected in series with the LED. The minimum power rating for the resistor can be determined using Ohm's Law also. The LED will have a voltage drop of approximately 1.7 volts. Because the resistor is connected in series with the LED, it will have a voltage drop of 10.3 volts. The power dissipation of the resistor can now be determined.

$$P = \frac{E^2}{R}$$

$$P = \frac{10.3^2}{620}$$

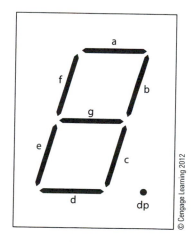

FIGURE 5-6 *Seven-segment display*

$$P = \frac{106.09}{620}$$

$$P = 0.171 \text{ watt}$$

A ¼-watt resistor can be employed in this circuit.

PHOTODIODES

The photodiode is so named because of its response to a light source. Photodiodes are housed in a case that has a window that permits light to strike the semiconductor material, Figure 5-7. Photodiodes can be used in two basic ways.

Photovoltaic

Photodiodes can be used as photovoltaic devices. When in the presence of light, they will produce a voltage in a manner similar to that of solar cells. The output voltage is approximately 0.45 volt. The current capacity is small, and use is generally limited to applications such as operating light metering devices. The basic schematic for a photodiode used as a photovoltaic device is shown in Figure 5-8. Note the symbol used to represent a photodiode. Arrows pointing toward the diode indicate that it must receive light to operate.

Photoconductive

Photodiodes can also be used as photoconductive devices. When used in this manner, they are connected reverse biased, Figure 5-9. In the presence of darkness, the amount of reverse current flow is extremely

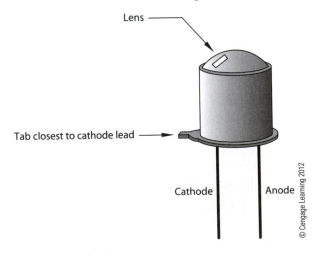

FIGURE 5-7 *Photodiode*

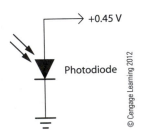

FIGURE 5-8 *Photodiode used as a photovoltaic device*

small, similar to that of a junction diode connected reverse biased. This current is referred to as the **dark current** (I_D) and is generally in the range of a few nanoamperes. For most practical purposes, dark current is generally considered to be zero.

When exposed to light, photons enter the depletion region and create electron–hole pairs, increasing conductivity in the reverse direction. The increased conduction may permit several milliamperes of current to flow. This is known as **light current** (I_L). The great advantage of the photodiode over other photoconductive devices, such as the cad cell, is speed of operation. Photodiodes can operate at very high frequencies.

LED Devices

Light-emitting diodes are used literally in hundreds of devices that range from simple indicating lights to high-intensity lighting for businesses. Many Christmas lights are composed of LEDs. They produce bright lights that consume about one-fourth the energy of conventional incandescent lights. One of the most common devices that employs the use of an LED and a photodiode is the optical mouse used with many computers, Figure 5-10. The light-emitting diode supplies light to the surface on which the mouse sits. As the mouse is moved, the photodiode detects the change of light intensity. The LED and photodiode are shown in Figure 5-11.

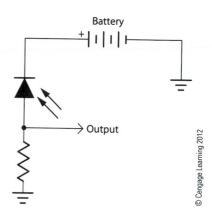

FIGURE 5-9 *Photodiode used as a photoconductive device*

FIGURE 5-10 *Optical mouse used with many computers*

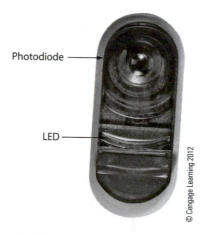

FIGURE 5-11 *The optical mouse contains both a light-emitting diode and a photodiode.*

UNIT 5 REVIEW QUESTIONS

1. What do the letters LED stand for? _____

2. If an LED is connected into an AC circuit, will the output voltage be AC or DC? _____

_____

3. What is the average forward voltage drop of an LED? _____

4. Refer to the symbol shown in Figure 5-12. What does the arrow pointing away from the diode
indicate? _____

FIGURE 5-12 *Light-emitting diode.*

5. An LED is to be connected into a 40-volt DC circuit. What size resistor should be used to limit the
current flow through the LED? _____

6. When used as a photovoltaic device, how much voltage is generally produced by a photodiode?

_____

7. Explain the difference between the schematic symbol used to indicate an LED and the symbol used
to indicate a photodiode. _____

_____

8. What is the greatest advantage of a photodiode used as a photoconductive device as compared to
other photoconductive devices such as a cad cell? _____

_____

_____

UNIT 6

SINGLE-PHASE RECTIFIERS

OBJECTIVES

After studying this unit, the student should be able to:

- Discuss the operation of single-phase rectifiers.

- Construct a half-wave rectifier.

- Connect a two-diode-type full-wave rectifier.

- Connect a bridge rectifier.

- Compute the output voltage for different types of rectifiers.

Rectifiers are devices that convert alternating or bidirectional current flow into direct or unidirectional current flow. Rectifiers can be divided into two basic types: single phase and three phase. Once the rectifier has changed the AC current into DC current, filters are used to smooth the pulsations and produce a smoother and more constant output voltage.

TYPES OF SINGLE-PHASE RECTIFIERS

Half-Wave Rectifier

The simplest of all rectifiers is the single-phase half-wave. This rectifier is constructed by connecting a single diode in series with a load connected to an AC line, Figure 6-1. Because the diode operates like an electrical check valve, it will permit current to flow in only one direction. When the polarity of the AC voltage applied to the diode is such that the anode terminal is more positive than the cathode, the diode is forward biased and current will flow through it. When the polarity of the AC voltage changes so that the anode lead of the diode is more negative than the cathode, the diode is reverse biased and no current can flow. The half-wave rectifier derives its name from the fact that it permits current to flow during half the AC waveform and turns off during the other half.

Note the pulsations of the rectified DC voltage shown in Figure 6-1. This illustration shows that the voltage turns on and off at regular intervals but never changes polarity. These pulsations are

FIGURE 6-1 *A single-phase half-wave rectifier can be constructed with one diode connected into an AC line.*

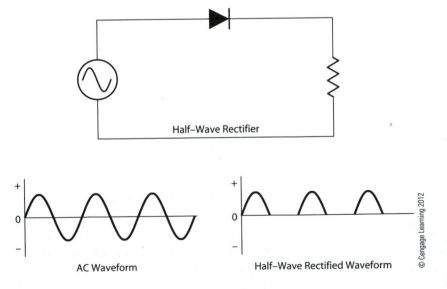

Half–Wave Rectifier

AC Waveform

Half–Wave Rectified Waveform

© Cengage Learning 2012

To Determine the Average Value for a Half-Wave Rectifier		
Peak	Multiply by:	0.3185
rms	Multiply by:	0.45

FIGURE 6-2 *The average voltage value for a single-phase half-wave rectifier is 0.3185 of the peak value or 0.45 of the rms value.*

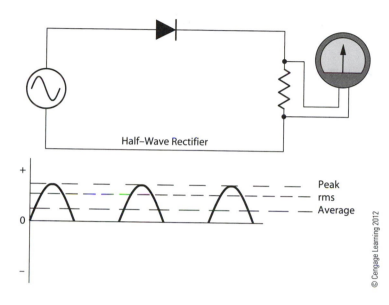

Half–Wave Rectifier

Peak
rms
Average

© Cengage Learning 2012

generally referred to as **ripple.** The amount of ripple is determined by the type of rectifier employed. The single-phase half-wave rectifier has the most ripple of any rectifier.

The amount of output DC voltage is referred to as the average voltage. A DC voltmeter connected across the load resistor in Figure 6-2 would indicate the average voltage value. The average DC voltage value can be calculated from either the peak value or the rms value of the AC waveform. The peak voltage value is the highest value of voltage reached by the waveform. The rms value is the effective heating value. Most listed AC voltages are rms, unless otherwise stated. Most AC voltmeters measure the rms value.

The average DC voltage value for a half-wave rectifier can be 60, determined by multiplying the peak AC voltage by 0.3185 or the rms AC voltage by 0.45 and then subtracting the diode voltage drop of about 0.7 volt.

EXAMPLE: Assume a half-wave rectifier is used to supply a DC voltage to a load. Also assume that the AC source voltage is 24 volts. How much DC voltage will be applied to the load?

SOLUTION 1: The rms value of AC voltage is 24 volts. Multiply the rms value by 0.45 and then subtract a voltage drop of 0.7 volt for the diode.

$$(24 \times 0.45) - 0.7 = 10.1 \text{ VDC}$$

SOLUTION 2: Change the rms value to the peak value and then multiply by 0.3185. Then subtract the diode voltage drop.

$$24 \times 1.414 = 33.936 \text{ volts peak}$$

$$(33.936 \times 0.3185) - 0.7 = 10.1 \text{ VDC}$$

Single-Phase Full-Wave Rectifiers

There are two types of single-phase full-wave rectifiers: the two-diode type and the bridge type. The two-diode type requires a center-tapped transformer, Figure 6-3. To understand the operation of the rectifier, assume that the transformer shown in Figure 6-4 has a secondary voltage of 24 VAC measured between points A and B. Because the secondary winding is center-tapped, a voltage of 12 VAC can be measured from A to C or from B to C. Now assume that at this point in time, point A is 24 volts positive with respect to point B.

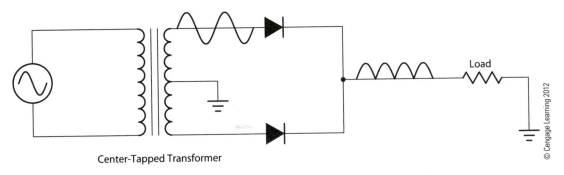

Center-Tapped Transformer

FIGURE 6-3 *Single-phase two-diode-type full-wave rectifier*

FIGURE 6-4 *Point A is more positive than point B.*

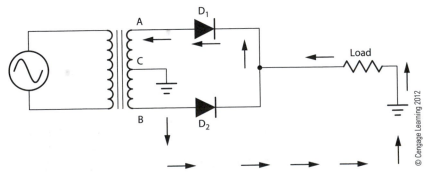

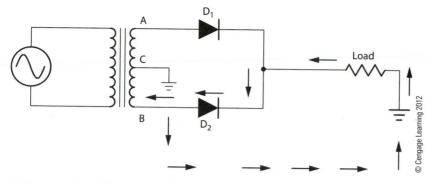

FIGURE 6-5 *Point B is more positive than point A.*

The center-tap is, therefore, 12 volts more negative than point A and 12 volts more positive than point B. Current will always move from the more negative to the more positive terminal. Current cannot flow from point B because diode D_2 is reverse biased. Current can flow from the center-tap, through ground to the load resistor, and back to point A through diode D_1. Note that current did flow through the load resistor during this half cycle of AC voltage.

During the next half cycle, point B becomes more positive than point A, Figure 6-5. Current cannot flow from point A because diode D_1 is now reverse biased. Current will now flow from the transformer center-tap, through ground to the load resistor, and return to point B through diode D_2. Note that current again flowed through the load resistor during this half cycle. This rectifier changes both halves of the AC waveform into DC.

The average DC voltage for a single-phase full-wave rectifier is 0.637 of the peak value or 0.9 of the rms value. When using a center-tapped transformer, however, you must consider that the rectifier is actually operating across half the secondary winding. In this example, the center-tap voltage is 12 VAC. Use this value to determine the average DC voltage.

$$12 \text{ VAC} \times 1.414 = 16.968 \text{ volts peak}$$

$$(16.968 \times 0.637) - 0.7 = 10.109 \text{ VDC}$$

or

$$(12 \text{ VAC} \times 0.9) - 0.7 = 10.1 \text{ VDC}$$

Single-Phase Bridge-Type Rectifiers

Bridge-type rectifiers require the use of four diodes instead of two. They are also considered to be less efficient than the two-diode-type full-wave rectifier because during each half cycle the voltage must be dropped across two diodes in series instead of one. The bridge-type

FIGURE 6-6 *Bridge-type rectifiers require four diodes.*

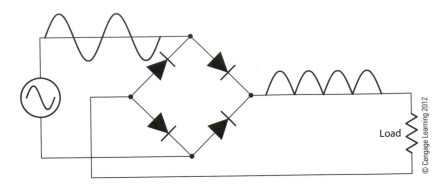

FIGURE 6-7 *Point X is more positive than point Y.*

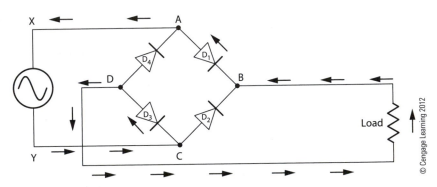

rectifier, however, has an advantage in that it does not require the use of a center-tapped transformer, Figure 6-6.

To understand the operation of the bridge-type rectifier, refer to the circuit shown in Figure 6-7. In this circuit, it is assumed that at this point in time, point X of the AC power supply is more positive than point Y. Current will flow from point Y of the power supply to point C of the bridge circuit. At this point, diode D_2 is reverse biased and will not conduct. Diode D_3, however, is forward biased. Current will flow through diode D_3 to point D of the bridge rectifier. At this point, diode D_4 is reverse biased and will not conduct. Current flows from point D of the bridge, through the load to point B. At this point, both diodes D_1 and D_2 are forward biased. Current, however, will not flow through diode D_2 because point C is more negative than point A. The current, therefore, flows through diode D_1 to point A and then to point X of the power supply. Note that current did flow through the load during this half cycle of AC voltage.

During the next half cycle, point Y becomes more positive than point X, Figure 6-8. Current will now flow from point X of the AC power supply to point A of the bridge rectifier. Diode D_1 is reverse biased and will not conduct. Current will flow through diode D_4 to point D of the bridge. Diode D_3 is now reverse biased and will not conduct.

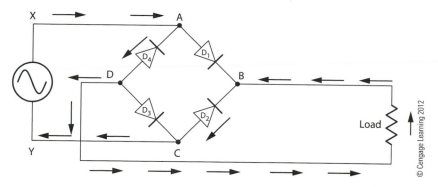

FIGURE 6-8 *Point Y is more positive than point X.*

Current must now flow through the load to point B of the bridge rectifier. Diodes D_1 and D_2 are again forward biased, but during this half cycle point Y is more positive than point X. Current will flow through diode D_2 and return to point Y of the power supply. A photograph of a bridge rectifier contained in a single case is shown in Figure 6-9.

The average DC voltage for the bridge rectifier can be computed in the same manner as the two-diode-type full-wave rectifier. Multiply the peak voltage value by 0.637 or the rms voltage value by 0.9. The bridge rectifier, however, has two diodes connected in series during each half cycle. The voltage drop of both diodes must be subtracted in the calculation.

FIGURE 6-9 *Bridge rectifiers in a single package*

UNIT 6 REVIEW QUESTIONS

1. What is a rectifier? _____

_____

2. What is the simplest type of rectifier? _____

_____

3. What type of single-phase rectifier requires the use of a center-tapped transformer? _____

_____

4. What type of single-phase rectifier requires the use of four diodes? _____

_____

5. An 18-volt center-tapped transformer is used to construct a two-diode-type rectifier. What will the DC output voltage be? _____

_____

6. What is the most efficient type of single-phase rectifier? _____

_____

7. A bridge rectifier is connected to the output of a 36-volt transformer. What will be the DC output voltage? _____

_____

UNIT 7

THE POLYPHASE RECTIFIER

OBJECTIVES

After studying this unit, the student should be able to:

- Discuss the operation of a three-phase rectifier.

- Compute the average DC output voltage of a half-wave, three-phase rectifier.

- Compute the average DC voltage on a full-wave, three-phase rectifier.

- Connect a half-wave, three-phase rectifier using discrete components.

- Connect a full-wave, three-phase rectifier using discrete components.

In the previous unit, single-phase AC voltage was changed into DC voltage. Industry, however, is operated by three-phase power, so it is necessary to change three-phase alternating current into direct current. There are two types of three-phase rectifiers: the half-wave and the bridge.

THE HALF-WAVE RECTIFIER

The half-wave, three-phase rectifier must be connected to a wye (star) system that has a grounded center tap or a fourth conductor connected to the center tap, Figure 7-1.

Notice in Figure 7-1 that a diode is connected in series with each phase of the system. The diodes are forward biased when the voltage of each line becomes positive and reverse biased when the voltage becomes negative. As the voltage of each of the three-phase lines goes positive, current flows through the load resistor to ground. It then flows from ground to the center tap of the transformer to complete the circuit. This rectifier has a higher average voltage output and less ripple than a single-phase, full-wave rectifier.

Although the figure shows a half-wave rectifier, it is changing three separate phases that are 120 degrees out of phase with each other into direct current. The rectified DC voltage never falls back to 0 volt before another of the three-phase lines begins conducting. Therefore, the DC voltage never reaches the 0 reference line as it does with a single-phase, full-wave rectifier. Figures 7-2(A) and 7-2(B) show the difference between these two rectifiers.

THE BRIDGE RECTIFIER

The three-phase bridge rectifier is often used for industrial applications because it has less ripple than the half-wave rectifier and does not require a center-tapped, wye-connected transformer for operation, Figure 7-3. It needs only to be connected to three-phase power for operation. Therefore, power can be supplied by either a wye or delta system.

FIGURE 7-1 *Three-phase, half-wave rectifier*

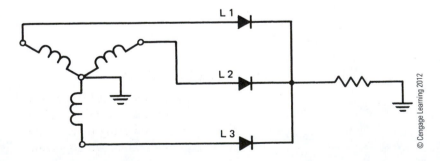

L 1

L 2

L 3

© Cengage Learning 2012

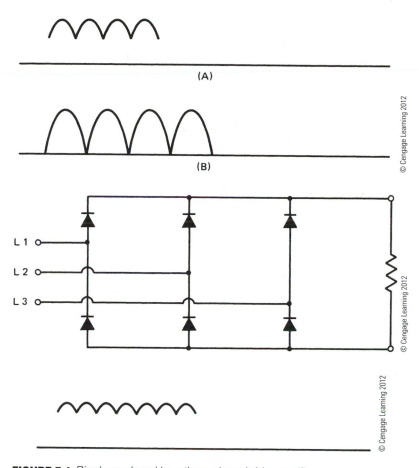

(A)

(B)

FIGURE 7-2 *(A) Ripple produced by a three-phase, half-wave rectifier (B) Ripple produced by single-phase, full-wave rectifier*

FIGURE 7-3 *Three-phase bridge rectifier*

L 1
L 2
L 3

FIGURE 7-4 *Ripple produced by a three-phase bridge rectifier*

The three-phase bridge-type rectifier also has a higher average DC voltage than the three-phase, half-wave rectifier because the bridge rectifier changes both the positive and negative halves of the AC voltage into DC, Figure 7-4.

A very common application of this rectifier is in the alternator of an automobile. Most automobiles use a wye-connected, three-phase alternator to supply the power of the charging system. The three-phase AC power produced by the alternator is converted into DC by a three-phase bridge rectifier, Figure 7-5.

AVERAGE VOLTAGE CALCULATIONS

The average output voltage for a three-phase rectifier is higher than that for a single-phase rectifier. The reason for this is that there is less ripple when a three-phase rectifier is used. When using three-phase rectifiers, the average DC voltage output is higher than the rms value

FIGURE 7-5 *Automobile alternator circuit*

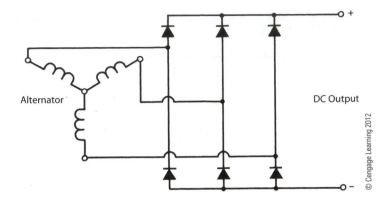

of the AC voltage. This is due to the fact that when three-phase voltage is rectified, the waveform never drops back to 0 volt, as shown in Figures 7-2A and 7-4. When a three-phase, half-wave rectifier is used, the average DC voltage can be computed by multiplying the peak value of voltage by 0.827, or by multiplying the rms value of AC voltage by 1.169. For example, compute the DC average voltage if 480 volts three-phase is rectified by a three-phase, half-wave rectifier.

SOLUTION:

In the first method, the peak value of voltage is found by multiplying the rms value by 1.414.

$$480 \times 1.414 = 678.72 \text{ volts}$$

The peak voltage is now multiplied by 0.827 to find the average DC voltage.

$$678.72 \times 0.827 = 561.3 \text{ volts}$$

The average DC voltage can also be computed by multiplying the rms value by 1.169.

$$480 \times 1.169 = 561.12 \text{ volts}$$

The slight difference in answers is caused by rounding off values.

The average value of DC voltage can be computed for a full-wave, three-phase rectifier by multiplying the peak value of voltage by 0.955, or by multiplying the rms value of AC voltage by 1.35. In the following example, the average DC output voltage for a three-phase, full-wave rectifier connected to 480 volts AC is computed.

In the first method, the peak voltage will be multiplied by 0.955.

$$678.72 \times 0.955 = 648.12 \text{ volts}$$

The average voltage will now be computed using the rms value of AC voltage.

$$480 \times 1.35 = 648 \text{ volts}$$

APPLICATIONS

Commercial and industrial applications sometimes require a direct current with very low ripple. Some electronic equipment, such as radio and television transmission equipment, must have a very smooth direct current to operate. The ripple produced by changing AC voltage into DC voltage can be made much smoother by the use of filters. These will be discussed further in Unit 8. The amount of filtering required is partly determined by the amount of ripple the rectifier produces. The three-phase bridge rectifier produces much less ripple than the single-phase bridge rectifier. The three-phase bridge rectifier requires much less filtering, therefore, than a single-phase bridge rectifier of the same power rating.

The amount of ripple produced by a rectifier can become a matter of great importance. For this reason, there are rectifiers that operate on 6, 12, or 18 phases. These multiple phases are obtained by various transformer connections to a three-phase system. When a multiphase rectifier is used, it is generally connected as a half-wave rectifier. A six-phase, half-wave rectifier has the same ripple as a three-phase bridge rectifier, Figure 7-6.

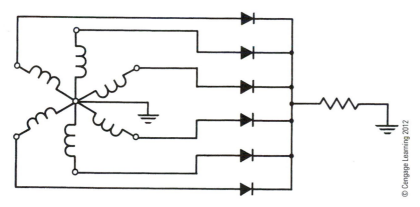

FIGURE 7-6 *Six-phase, half-wave rectifier*

© Cengage Learning 2012

UNIT 7 REVIEW QUESTIONS

1. What type of three-phase rectifier requires the use of a wye-connected transformer with a center tap? _____

_____

2. How many diodes are required to construct a three-phase, full-wave rectifier? _____

_____

3. Which type of rectifier has less ripple, a full-wave, single-phase rectifier or a half-wave, three-phase rectifier? _____

_____

4. A three-phase, half-wave rectifier is connected to a 208-volt, wye-connected system. What is the average DC voltage produced by the rectifier? _____

_____

5. A three-phase, full-wave rectifier is connected to a 560-volt, delta-connected system. What is the average DC voltage output? _____

_____

FILTERS

OBJECTIVES

After studying this unit, the student should be able to:

- Discuss the operation of filters in an electronic circuit.
- Discuss the differences between capacitive and inductive filters.
- Connect a capacitive filter into a circuit.
- Connect an inductive filter into a circuit.

In previous units, AC voltage was changed into DC voltage with a rectifier. Regardless of the type of rectifier used, the DC voltage pulsates. The amount of pulsation (ripple) is determined by the type of rectifier used. A half-wave, single-phase rectifier produces the greatest amount of ripple. The three-phase bridge rectifier produces the least. Some types of loads such as DC motors, DC relays, and battery chargers will operate without problems on pulsating (unfiltered) DC, but other electronic loads will not.

PUTTING THE FILTER TO USE

Assume that a radio or tape player that is designed for use in an automobile is operated inside a house. The power in a home is 120 VAC, and the power in an automobile is 12 VDC. The 120 VAC will have to be stepped down to 12 VAC with a transformer, and then rectified into DC, Figure 8-1.

Although 12 VDC is now available to operate the radio, it will probably produce an annoying hum. The hum is caused by the pulsations of the DC voltage being applied to the radio. If the radio is to operate properly, the pulsations must be removed.

The pulsations or ripple of the DC voltage can be removed with the proper filter. There are two components used for filtering: a **capacitor** is used to filter the voltage, and an **inductor** or **choke** to filter the current.

The Capacitor as a Filter

When a capacitor is used as a filter, it is connected in parallel with the output of the power supply, Figure 8-2. Assume that the rectifier or power supply produces an output voltage that has a peak value of 20 volts, Figure 8-3.

FIGURE 8-1 *DC power supply*

FIGURE 8-2 *Capacitor filter*

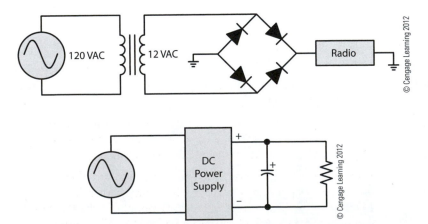

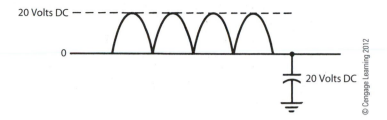

20 Volts DC

0

20 Volts DC

© Cengage Learning 2012

FIGURE 8-3 *Unfiltered DC with a peak value of 20 volts*

As the voltage rises to its peak value of 20 volts, the capacitor charges to 20 volts also. When the DC waveform tries to drop back toward zero, the line voltage becomes less than the 20 volts the capacitor has been charged to. Therefore, the capacitor discharges back into the line in an effort to keep the voltage from decreasing in value, Figure 8-4. The amount of filtering that is accomplished is determined by two things:

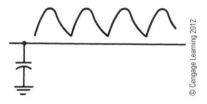

FIGURE 8-4 *Small amount of filtering*

- The amount of current the power supply must furnish to the load
- The amount of capacitance connected to the circuit

If the power supply is to furnish only a few milliamps of current, only a small amount of capacitance is needed to filter the voltage. If the power supply must furnish several amps, however, several thousand microfarads (μF) of capacitance will be needed.

Computing Filter Effectiveness

The output voltage of a full-wave (or any other) rectifier is called the average value and is equal to the peak value of the waveform multiplied by 0.637. For instance:

1. Assume an AC voltage of 24 volts (rms) is changed into DC.

2. The peak value of the waveform is 24 × 1.414 = 33.9 VAC.

3. The output voltage of the rectifier is 33.9 × 0.637 = 21.6 VDC. (The average value of voltage is 21.6 volts.)

4. When the output voltage of a rectifier is filtered, the voltage increases in value. The amount of increase is in proportion to the amount of filtering.

5. The waveform shown in Figure 8-5(A) has been filtered only a small amount. The output voltage of this power supply is only slightly higher than an unfiltered power supply.

6. If more capacitance is added to the circuit, the waveform will begin to appear more like the waveform shown in Figure 8-5(B). The voltage is much higher above the 0 reference line than the waveform of Figure 8-5(A). The output

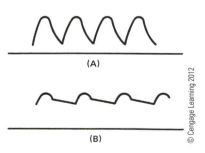

(A)

(B)

FIGURE 8-5 *(A) Small amount of capacitance. (B) Larger amount of capacitance*

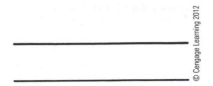

FIGURE 8-6 *DC voltage completely filtered*

voltage of the power supply is higher because the voltage doesn't drop as close to 0 volt.

7. If enough capacitance is added to the circuit to completely filter the ripple, the output waveform appears as a straight line, Figure 8-6.

When all the ripple has been filtered out of the DC voltage, the output voltage of the power supply will be the same value as the peak voltage of the AC applied voltage being rectified. Assume that 24 volts AC rms is connected to a full-wave rectifier. The peak value of the AC waveform is 33.9 volts. If the DC rectified voltage is filtered so that it has no ripple, the DC output voltage will be 33.9 volts.

The Choke Coil as a Filter

The choke coil does for current what the capacitor does for voltage. It is connected in series with the output of the rectifier, Figure 8-7.

When the current begins to rise from 0 toward its peak value, a magnetic field is created around the choke coil, Figure 8-8. As the current reaches its peak value, the magnetic field does also. They decrease from their peak value back toward 0 in the same way, Figure 8-9. As the magnetic field collapses around the choke coil, it induces a current back into the coil. This induced current flows in the same direction as the circuit current. Therefore, it aids the circuit current by trying to keep a continuous current flow throughout the circuit. The size of the choke coil needed to filter the current is determined by the amount of power the circuit must produce.

FIGURE 8-7 *Connection of filter choke*

FIGURE 8-8 *Magnetic field increasing with an increase of current*

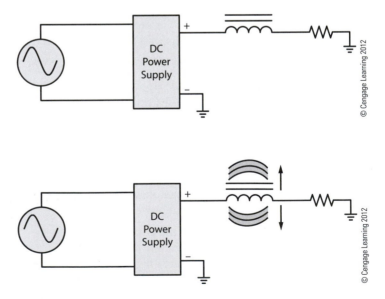

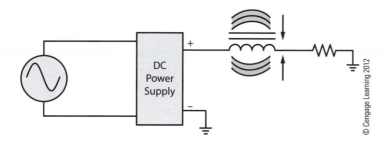

FIGURE 8-9 *Magnetic field inducing current back into the coil*

COMPARISON OF THE FILTERS

There are various ways to connect capacitors and chokes into a circuit for filtering. A power supply designed for low-power application may use a single capacitor filter. A common filtering arrangement is the "PI" filter. It gets its name because it resembles the Greek letter "PI," Figure 8-10.

Power supplies designed to deliver a larger current generally use a choke input filter, Figure 8-11. Choke input means that the choke is connected ahead of the capacitor in reference to the rectifier. This is generally done to keep the surge of current down when the power is turned on. Recall that:

- A discharged capacitor looks like a short circuit to ground when power is first applied to it.

- This can cause a huge current spike, which can destroy the diodes in the rectifier.

- By connecting the choke coil ahead of the capacitor, the rate at which the current can rise is limited.

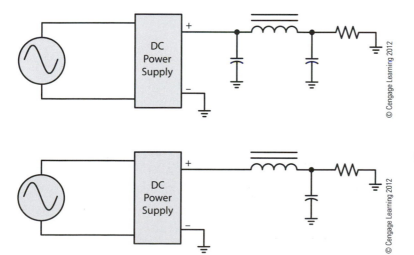

FIGURE 8-10 *"PI" filter*

FIGURE 8-11 *Choke input filter*

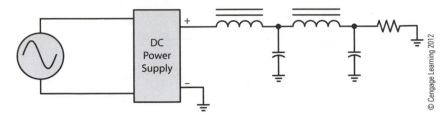

FIGURE 8-12 *Choke input "PI" filter*

- An inductor opposes a change of current. This opposition to a change of current prevents the high current spike when the power is first turned on.

A common industrial choke input filter is shown in Figure 8-12. Notice that this is a "PI" filter with the addition of a choke input.

The schematic drawings of filter networks may show one capacitor connected in parallel with the output of the rectifier. This capacitor on the schematic may actually be several capacitors connected in parallel with each other to increase the total capacitance.

UNIT 8 REVIEW QUESTIONS

1. What type of rectifier produces the greatest amount of ripple? _____

_____

2. What type of rectifier produces the least amount of ripple? _____

_____

3. What electronic component is used to filter voltage? _____

_____

4. What electronic component is used to filter current? _____

_____

5. How is a capacitor filter connected into the circuit? _____

_____

6. How is an inductor or choke connected into the circuit? _____

_____

7. What is the advantage of the choke input filter? _____

_____

8. A full-wave bridge rectifier is connected to a 36-volt transformer. If enough capacitance is used to completely filter the ripple, what will be the output voltage of the rectifier? _____

_____

UNIT 9

SPECIAL DIODES

OBJECTIVES

After studying this unit, the student should be able to:

- Discuss the operation of a zener diode.
- Draw the schematic symbol for a zener diode.
- Construct a voltage regulator using a zener diode.
- Discuss the operation of a tunnel diode.
- Draw the schematic symbol for a tunnel diode.
- Discuss varactor diodes.
- Draw the schematic symbol for a varactor diode.
- Discuss the operation of a Shockley diode.
- Draw the schematic symbol for a Shockley diode.
- Discuss the operation of a Schottky diode.
- Draw the schematic symbol for a Schottky diode.
- Discuss the operation of an IMPATT diode.
- Discuss Gunn diodes.
- Discuss the operation of a PIN diode.

In previous units, rectifier or junction diodes and light-emitting diodes have been discussed. Although junction diodes comprise a large portion of the diodes used throughout industry and in everyday electronic devices, other types of diodes intended to perform a specific task do exist. Some of these special diodes will be discussed in this unit.

ZENER DIODES

Zener diodes are special diodes that have a low reverse voltage breakdown rating. They are named for C. A. Zener, who analyzed the voltage breakdown of electrical insulators. Zener diodes generally have a reverse breakdown voltage of about 3–200 volts. They are intended to be operated with current flowing through them in the reverse direction. The symbol for a zener diode is shown in Figure 9-1.

When the reverse breakdown voltage of a diode is exceeded, the current suddenly begins to flow with almost no restriction. This is generally referred to as an avalanche condition. Zener diodes are sometimes referred to as avalanche diodes. This avalanche current is referred to as the **zener region**, Figure 9-2. Note in Figure 9-2 that the reverse voltage drop is relatively constant over a wide range of current. Diodes are not harmed when they

FIGURE 9-1 *Zener diode symbol*

FIGURE 9-2 *The current avalanches in the reverse direction.*

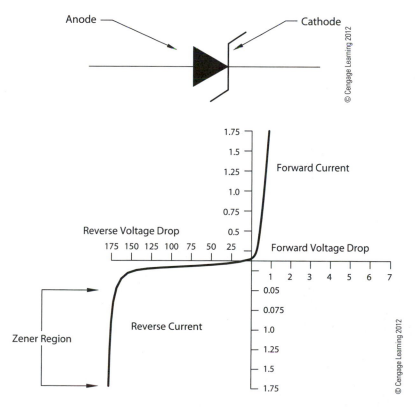

FIGURE 9-3 *Zener diode used as a voltage regulator*

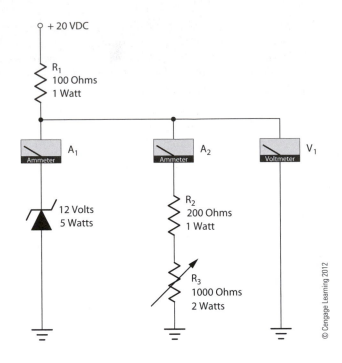

operate in the zener region as long as their electrical values of current or watts are not exceeded.

Because the voltage drop across the zener diode is constant, the diode is generally used as a voltage regulator. Any device or devices connected in parallel with the zener will have the same voltage applied. An example circuit is shown in Figure 9-3. The zener diode has a reverse voltage rating of 12 volts and a power rating of 5 watts. Resistor R_1 is used to limit the current in the circuit. Resistors R_2 and R_3 represent the load for the circuit. Note the supply voltage must be greater than the zener voltage for the circuit to operate.

Resistor R_1 and the zener diode form a series circuit to ground. Because the zener has a constant voltage drop of 12 volts, resistor R_1 will have a voltage drop of 8 volts (20 volts − 12 volts = 8 volts). Resistor R_1, therefore, will permit a total current flow of 0.08 A (80 mA) in this circuit (8 volts/100 Ω = 0.08 A).

The load circuit, R_2 and R_3, is connected in parallel with the zener diode. Because the load is in parallel with the zener, the voltage dropped across the zener will be applied to the load. If the zener maintains a constant voltage drop of 12 volts, the load will have a constant applied voltage of 12 volts. The maximum current that can flow through the load is 0.06 A (60 mA), (12 volts/200 Ω = 0.06 A). Note that the value of R_1 was chosen to ensure that there would be sufficient current to operate the load. In order for a load current of

60 mA to flow, the value of R_3 is adjusted to 0 Ω. At this point, ammeter A_2 would indicate a flow of 60 mA and ammeter A_1 would indicate 20 mA. The values of A_1 and A_2 will always add to equal the maximum current that can flow in the circuit, (60 mA + 20 mA = 80 mA). Voltmeter V_1 will indicate a value of 12 volts.

Now assume that resistor R_3 is adjusted to a value of 200 ohms. The load now has a resistance of 400 Ω (200 Ω + 200 Ω = 400 Ω). Ammeter A_2 will now indicate a current flow of 30 mA (12 volts/400 Ω = 0.030). Ammeter A_1, however, now indicates a current flow of 50 mA (50 mA + 30 mA = 80 mA). Voltmeter V_1 indicates a value of 12 volts.

Zener diodes can be tested in the same manner as rectifier diodes provided the ohmmeter voltage is less than the reverse voltage value of the zener. For example, an ohmmeter that provided an output voltage of 6 volts could not be used to test a zener diode with a reverse voltage rating of 5 volts. The zener diode would appear to be shorted even if it were not.

Zener Diode Application

A good example of how a zener diode can be used as a voltage regulator can be found in the charging circuit of many motorcycles, Figure 9-4. In this circuit, the alternator is used to produce the direct current needed to operate the electrical system of the motorcycle and charge the battery. When a 12-volt battery is fully charged, it will exhibit a voltage of 14 volts across its terminals. If the voltage supplied to the battery becomes greater than 14 volts, there is danger of overcharging the battery.

In this circuit, the alternator is connected in parallel with the battery. A power zener diode is connected in parallel with both the alternator and battery. Because electrical components connected in parallel must have the same voltage, the zener diode will not permit the voltage supplied to the battery to become greater than 14 volts. Notice that there is no current limiting resistor connected in series with the zener diode. The current in this circuit is limited by the internal impedance of the alternator.

FIGURE 9-4 *Typical voltage regulator circuit for a motorcycle*

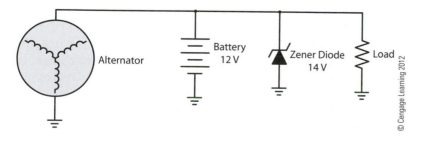

© Cengage Learning 2012

FIGURE 9-5 *Tunnel diodes exhibit a negative resistance.*

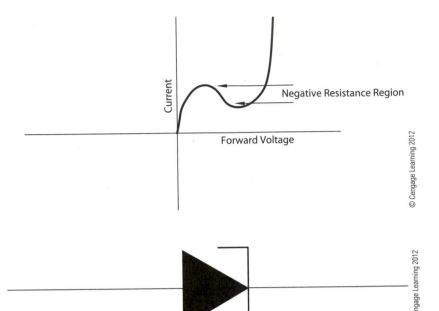

FIGURE 9-6 *Tunnel diode symbol*

TUNNEL DIODES

Tunnel diodes are often referred to as **Esaki** diodes. Esaki is the man who discovered that heavy doping caused a tunneling effect of charged current carriers through the depletion zone at the diode junction. Tunnel diodes exhibit a negative resistance for a specific range of forward voltage, Figure 9-5. The negative resistance characteristic causes the current to decrease with an increase of voltage. This negative resistance characteristic permits the tunnel diode to be operated as an oscillator or as an amplifier, although it has only two leads. They are generally low-power devices that can be operated at microwave frequencies. Although the tunnel diode can operate at microwave frequencies, it is relatively free from radiation effects. A schematic symbol of a tunnel diode is shown in Figure 9-6.

VARACTOR DIODES

The varactor diode is also called the varicap diode or capacitive diode. Like the zener diode, the varactor diode is operated reverse biased. When reverse voltage is applied to the varactor diode, the junction exhibits a capacitance because of the separated charges in the depletion zone. The capacitance values are in the picofarad range. The most important feature of the varactor diode is that the capacitance value is controlled by the amount of reverse voltage applied to the device. The voltage/capacitive relationship for a typical varactor is shown in Figure 9-7. The schematic symbol for a varactor diode is shown in Figure 9-8.

A typical application for a varactor is shown in Figure 9-9. In this circuit, the varactor is used to control the operating frequency of an electronic tuner. The output of the oscillator is controlled by the LC (inductive–capacitive) tuned circuit. Because the varactor constitutes part of the capacitance for the circuit, a change in the DC voltage will affect the operating frequency of the oscillator.

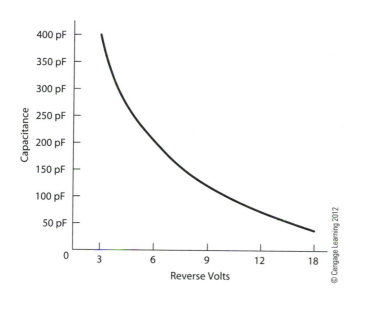

FIGURE 9-7 *Varactor diodes change capacitance with a change of voltage.*

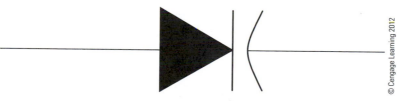

FIGURE 9-8 *Schematic symbol for a varactor diode*

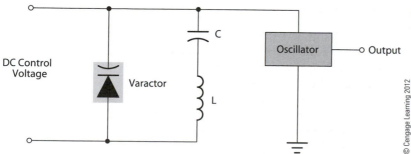

FIGURE 9-9 *Varactor used to control an electronic tuner*

SHOCKLEY DIODES

Shockley diodes are very similar to tunnel diodes in that they have a negative resistance characteristic. They are actually classified as a **thyristor** because they operate in only two states, on or off. The Shockley diode is constructed by joining four semiconductor layers together to form a pnpn junction, Figure 9-10. The Shockley diode exhibits a very high resistance, like that of an open switch, until the voltage across it reaches a point called the breakover voltage. When breakover voltage is reached, the diode suddenly exhibits a very low resistance and begins to conduct, as in Figure 9-11. Once the Shockley diode begins to conduct, the voltage across it will suddenly drop to about 1 volt. Once conduction begins, it will continue to conduct until the current falls below a certain level called the holding current level.

The negative resistance characteristic permits the Shockley diode to be used as a simple relaxation oscillator, Figure 9-12. In this

FIGURE 9-10 *The Shockley diode is a PNPN junction.*

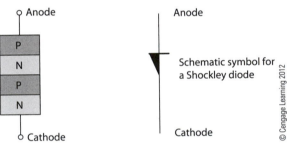

FIGURE 9-11 *Characteristic curve of a Shockley diode*

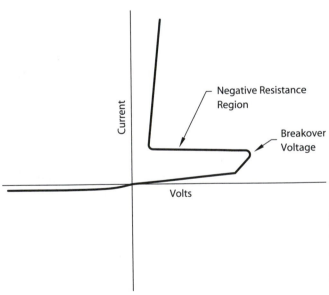

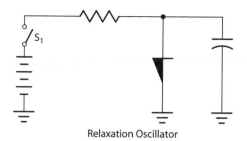

© Cengage Learning 2012

FIGURE 9-12 *Shockley diode used as a relaxation oscillator*

Relaxation Oscillator

Relaxation Oscillator Waveform

example circuit, the battery voltage is greater than the breakover voltage for the Shockley diode, but the value of the series resistor is high enough to limit the maximum circuit current to a value that is below the diode's holding current level. When switch S_1 is closed, the capacitor will begin to charge through the series resistor. When the voltage across the capacitor reaches the breakover value for the Shockley diode, it will turn on and discharge the capacitor. When the current level falls below the holding current value, the diode will turn off and the capacitor will begin charging again.

SCHOTTKY DIODES

Schottky diodes are also known as hot-carrier diodes. They are primarily used in high-frequency and high-speed switching applications. The Schottky diode is not constructed by joining two pieces of semiconductor material together, but rather by joining a piece of semiconductor material to a piece of metal, Figure 9-13. The semiconductor material is generally N-type, and the metal can be gold, silver, or platinum.

Because Schottky diodes have only majority current carriers in the semiconductor region and a conductor in the other, they have the ability to give up excess energy very rapidly. This permits them

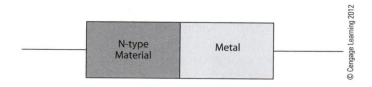

© Cengage Learning 2012

FIGURE 9-13 *Schottky diodes are constructed by joining a semiconductor material with a metal.*

FIGURE 9-14 *Schematic symbol for a Schottky diode*

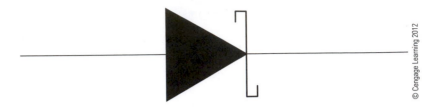

to switch on and off very quickly. Schottky diodes are generally used to emit high-frequency signals. The schematic symbol for a Schottky diode is shown in Figure 9-14.

PIN DIODES

PIN diodes are constructed by separating two semiconductor regions with an intrinsic region, Figure 9-15. Separating the p- and n-type regions with an intrinsic or undoped region produces some unique characteristics. If the PIN diode is connected reverse biased, it will act like an almost constant capacitance. When connected forward biased, it acts like a variable resistance. The resistance will decrease with an increase in forward current. PIN diodes are commonly used as a microwave switch by connecting DC to the diode. The variable resistance characteristics also permit it to be used as an attenuator. A high-frequency signal strength can be controlled by a varying DC bias. The schematic symbol for a PIN diode is shown in Figure 9-16.

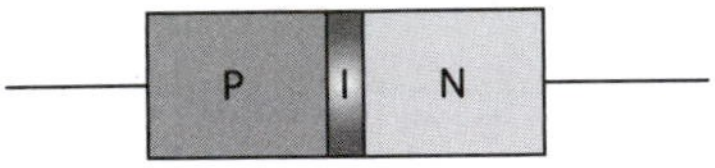

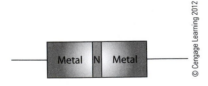

FIGURE 9-15 *PIN diodes are constructed by separating two semiconductor regions with an intrinsic (undoped) region.*

FIGURE 9-16 *Schematic symbol for a PIN diode.*

IMPATT DIODE

IMPATT is an acronym for impact avalanche and transit time diode. The IMPATT diode is a special diode used in microwave oscillators that operate between 10 and 100 GHz. Its name is derived from the way it operates. The IMPATT diode uses the delay time required for attaining an avalanche condition and the transit time necessary to produce a negative resistance characteristic.

GUNN DIODE

The Gunn diode is used in microwave oscillators. It is not a pn junction, but is constructed by joining two metal conductors on either side of an n-type gallium–arsenide semiconductor material, Figure 9-17. Like other microwave diodes, the Gunn diode exhibits a negative resistance characteristic.

FIGURE 9-17 *The Gunn diode is constructed by bonding a piece of n-type gallium–arsenide semiconductor material between two metal conductors.*

UNIT 9 REVIEW QUESTIONS

1. Explain the difference in operation between a zener diode and a common junction diode. _____

_____

2. What generally happens to a junction diode that is broken down into the zener region? _____

_____

3. What is the primary function a zener diode is used to perform? _____

_____

4. A 5-volt zener diode is to be connected to a 12-volt source. The current must be limited to a
maximum of 14 mA. What size current-limiting resistor should be connected in series with the
zener diode? _____

_____

5. A zener diode is to be used as a voltage regulator. Should the load be connected in series with the
zener diode or in parallel? _____

_____

6. What effect does heavy doping have on the tunnel diode? _____

_____

7. Should a varactor diode be connected forward or reverse biased? _____

_____

8. Why are Shockley diodes classified as thyristors? _____

_____

9. What type of diode is also known as a hot-carrier diode? _____

_____

10. How is the capacitance of a varactor diode controlled? _____

_____

UNIT 9 REVIEW QUESTIONS

11. How is a PIN diode constructed? _____

_____

12. How can a PIN diode be made to exhibit a constant capacitance? _____

_____

13. What type of diode can operate between 10 and 100 GHz? _____

_____

UNIT 10

THE TRANSISTOR

OBJECTIVES

After studying this unit, the student should be able to:

- Discuss the operation of a transistor.

- Name the two types of transistors.

- Find the parameters of a transistor in a semiconductor catalog.

- Test a transistor with an ohm-meter.

- Connect a transistor into an electronic circuit.

- Discuss the polarity connections for different types of transistors.

- Draw transistor symbols for both NPN- and PNP-type transistors.

The transistor is a semiconductor device made by joining together three layers of P- or N-type material. There are two basic types of transistors: the NPN and the PNP. The manner in which the layers of semiconductor material are joined together determines the transistor type, Figure 10-1.

Germanium and silicon are both used in the production of transistors, but silicon is used more often because of its ability to withstand heat. However, a silicon device requires more voltage to operate. Therefore, when low operating power is required, a germanium device may be employed.

EARLY TRANSISTOR HISTORY

Although silicon is now the most common semiconductor material used in the production of transistors, germanium was used to make the first transistor in 1947. Because of difficulty in refining silicon to an adequately pure form for use in semiconductor devices, germanium remained the dominant material until the mid-1960s. Early germanium transistors exhibited several undesirable characteristics. One day a piece of equipment would be operating properly, and the next it would die for no explainable reason. Electronic technicians and engineers coined the term *purple plague*. When a piece of equipment suddenly stopped operating, they would say, "The purple plague got it."

Another undesirable characteristic of germanium transistors is their susceptibility to thermal runaway. Thermal runaway is caused by a semiconductor's characteristic of decreasing resistance with an

FIGURE 10-1 *Transistors*

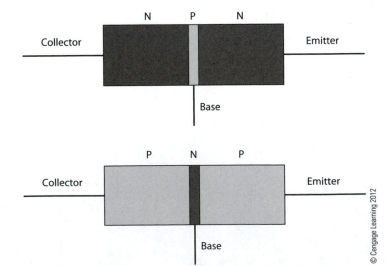

© Cengage Learning 2012

increase of temperature. When current flows through a semiconductor device, its temperature increases due to the resistance of the material. This increase in temperature causes the resistance to decrease, which permits more current to flow. Increased current flow produces increased temperature, which causes a further decrease of resistance and additional increase of current. This process continues until the device fails from overtemperature.

Silicon transistors attained popularity before the germanium purple plague was corrected, and although silicon exhibits a decrease of resistance with an increase of temperature just as germanium does, it is not as susceptible to thermal runaway.

IDENTIFYING A TRANSISTOR

Transistors are made with a wide variety of voltage, current, and power ratings. They are made with different sizes and styles of cases. They may also be in a plastic case or a metal case. The type of case generally determines the amount of power the transistor is designed to control, Figure 10-2.

Transistors can have:

- Voltage ratings from a few to several hundred volts
- Amperage ratings from milliamperes to 50 or more amperes
- Power ratings from milliwatts to several hundred watts

Additionally, some transistors are designed to operate at a few thousand cycles per second, whereas others can operate at several hundred million cycles per second.

Transistor ratings can vary greatly and without some means of identifying the transistor, its characteristics may never be known. The most commonly used system is the 2N registry. The transistors

Courtesy of International Rectifier Corp.

FIGURE 10-2 *Transistors shown in different case styles*

FIGURE 10-3 *Transistor specifications*

NUMBER	MATE-RIAL	POLA-RITY	POWER RATING PD	VOLTAGE RATING Vce	CURRENT RATING Ic	FREQUENCY
2N2222	S	N	0.5 W	30	150 mW	250 MHz
2N2907	S	P	1.8 W	40	150 mW	200 MHz
2N2928	G	P	150 mW	13	2 mW	400 MHz
2N6308	S	N	125 W	360	3 A	5 MHz
2N6280	S	N	250 W	140	20 A	30 MHz

© Cengage Learning 2012

listed in this system have a number preceded by 2N and can be found in a reference book.

Refer to Figure 10-3 and see that:

1. The number column gives the 2N number of the transistor.

2. The material column tells whether the transistor is made of germanium or silicon.

3. The polarity column indicates whether the transistor is an NPN or a PNP.

4. The power column tells the amount of heat that the transistor can dissipate. (Note: The 2N6308 and the 2N6280 transistors show power ratings of 125 and 250 watts, respectively. These power ratings can be obtained only if the transistors have been mounted on proper heat sinks and thermal compound used. The 2N2928 transistor has a power rating of 150 mW, which indicates 150 milliwatts, or 0.15 watt.)

5. The voltage rating column indicates the amount of voltage the device can withstand without breaking down.

6. The amperage column indicates the maximum amount of current the transistor can conduct. (Note: A transistor may not be able to hold off maximum voltage while conducting maximum current. The 2N6308 transistor has a power rating of 125 watts, a voltage rating of 360 volts, and a current rating of 3 amperes. If the transistor were to drop 350 volts while conducting 3 amperes of current, it would have to dissipate 1050 watts ($350 \times 3 = 1050$).)

7. The frequency column indicates the maximum operating frequency of the transistor. MHz means megahertz.

Some manufacturers, such as Motorola and RCA, use their own numbering system for some devices. Motorola often uses MJ, MJE, or SDT. RCA uses an SK number to identify some of their transistors. Some equipment manufacturers use their own special numbers to identify components. These numbers mean nothing to anyone except the manufacturer and in most cases cannot be identified.

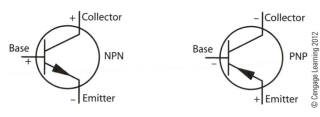

FIGURE 10-4 *Schematic symbols of transistors*

Transistor Schematics

Figure 10-4 shows the symbol for the NPN- and PNP-types of transistor and the polarity markings for each. Notice that the transistors have the base connected to the same polarity as the collector. The arrow on the emitter of the transistor points in the direction of conventional current flow, positive to negative.

OPERATION OF THE TRANSISTOR

The transistor operates like an electric faucet. The collector is the input and the emitter is the output of the device. The base is the control valve, Figure 10-5. The current flowing through the base and emitter controls the major current path, which is through the collector and emitter of the transistor. A few milliamperes of base current can control several hundred milliamperes of current through the collector and emitter circuit. The operation of the transistor will be covered in greater detail later in the unit.

Testing the Transistor

Transistors can be tested with an ohmmeter. The test will indicate whether the transistor is good or bad. If the polarity of the leads of the ohmmeter is known, it will indicate whether the transistor is NPN or PNP. To an ohmmeter, a transistor appears to be two diodes with their anodes (NPN) or their cathodes (PNP) connected.

As shown in Figure 10-6, an NPN transistor appears to the ohmmeter as two diodes that have their anodes connected. If the positive lead of the ohmmeter is connected to the base of the transistor,

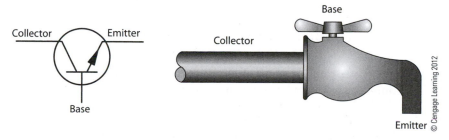

FIGURE 10-5 *Transistors operate like electric faucets.*

FIGURE 10-6 *Ohmmeter test*

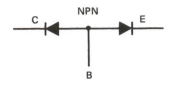

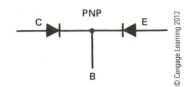

it shows continuity to both the collector and the emitter. If the negative lead of the ohmmeter is connected to the base of the transistor, it does not show continuity between the base and the collector or the base and the emitter.

The PNP transistor can be tested by reversing the polarity and connecting the negative lead of the ohmmeter to the base. The ohmmeter indicates continuity between the base and the collector and the base and the emitter. If the positive lead of the ohmmeter is connected to the base of the PNP transistor, it shows no continuity to the collector or the emitter.

The ohmmeter test is considered to be about 95 percent accurate, but there are some conditions under which the test would not be valid. If the transistor is breaking down under a high voltage, the ohmmeter will not be able to supply enough voltage to show the transistor defective. If the transistor is being broken down by heat, the ohmmeter will not supply enough power to show the transistor is bad.

Identifying the Leads

The leads of transistors packaged in standard TO3, TO5, or TO18 cases are relatively simple to identify, Figure 10-7. Transistors packaged in the TO92, TO220, or TO218 cases can sometimes be difficult because of the position of their leads, Figure 10-2. When testing a transistor in the TO5 or TO18 case, hold the transistor upside down. The three leads form a triangle on one side of the case. The little tab on the case is always closest to the emitter lead of the transistor, Figure 10-7(A). The TO3 case transistor leads can be identified by holding the transistor with the leads facing you and down. The lead to the left is the emitter and the lead to the right is the base. The collector is the case of the transistor. Refer to Figure 10-7(B). This can be seen more clearly in Figure 10-8, where the matching heat sink holes for the TO3 device are shown.

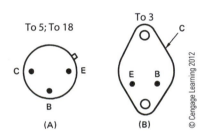

FIGURE 10-7 *Lead identification*

The plastic case transistors, such as the TO92, TO220, TO218, and so on, can have their leads at any position. The following procedure can be used to identify these transistors:

1. Assume the transistor is an NPN.

2. Assume that one lead of the transistor is the base. Connect the positive lead of the ohmmeter to it.

FIGURE 10-8 *A heat sink for TO3 case devices*

3. Touch each of the other two leads one at a time. If there is continuity to the other two leads, it is the base of the transistor and it is an NPN. If there is no continuity, assume another lead to be the base and repeat the test.

4. If none of the three transistor leads prove to be the base pin, assume the transistor to be a PNP. Repeat the procedure, using the negative lead of the ohmmeter connected to the base of the transistor.

5. If no lead can be identified as the base of the transistor, the component is probably not a transistor, or it is defective.

6. Once the base lead is identified and it is determined whether the transistor is NPN or PNP, the two remaining leads will have to be identified as to which is the collector and which is the emitter. (Note: The base of the transistor must have the same polarity as the collector in order for the transistor to operate, as shown in Figure 10-4.)

7. Assume that the transistor is an NPN and the base lead has been identified. Assume that one of the two remaining leads is the collector. Connect the positive ohmmeter lead to it. Connect the negative lead to the other pin. The ohmmeter should not indicate continuity.

8. Using a resistor of any value from 10 to 5000 ohms, touch one lead of the resistor to the positive ohmmeter lead and the base of the transistor to the other resistor lead. If the ohmmeter shows conduction, the assumption is correct and the positive ohmmeter lead is connected to the collector of the transistor. If the ohmmeter does not show continuity, reverse the leads and assume the other transistor pin to be the collector.

9. A PNP transistor can be tested in the same manner. Connect the negative ohmmeter lead to the collector and touch the resistor between the negative lead of the ohmmeter and the base pin of the transistor.

Current Flow

A transistor is not the same thing as a variable resistor. A transistor is a current-controlled device. The amount of base current controls the amount of current that flows through the collector and emitter. The transistor follows all the rules of Ohm's Law and its impedance changes, but the change of impedance is caused by a change of current flow through the device. In the circuit shown in Figure 10-9, resistor R_1 is used to limit the amount of current flow through the collector and emitter of the transistor. Resistor R_2 limits the base

FIGURE 10-9 *Basic transistor circuit*

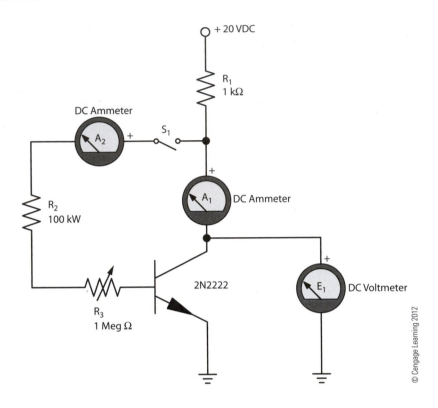

© Cengage Learning 2012

current if resistor R_3 is adjusted to 0 ohm. Resistor R_3 is used to control the amount of current flow to the base of the transistor.

If switch S_1 in Figure 10-9 is open, no current flows to the base of the transistor, and the transistor will not conduct current through the collector–emitter. In this condition, the transistor appears to be an open switch. The voltmeter will show a voltage drop across the collector–emitter of 20 volts. Because the transistor has a voltage drop of 20 volts with no current flow, the impedance of the transistor appears to be infinity ($20/0 = \infty$).

Assume that switch S_1 has been closed, and resistor R_3 has been adjusted to permit a current of 20 microamperes (μA) to flow through the base–emitter of the transistor. The 20 μA of base current permits a current of 2 mA to flow through the collector–emitter of the transistor. The voltmeter indicates a voltage of 18 V across the collector–emitter of the transistor. Because the transistor has a current of 2 mA flowing through it and a voltage drop of 18 V, it has an impedance of 9000 ohms (Ω) ($18/0.002 = 9000$).

If resistor R_3 is adjusted to permit a current flow of 40 μA through the base–emitter of the transistor, more current will be permitted to flow through the collector–emitter. Assume a collector–emitter current of 4 mA. The voltage drop across the collector–emitter of the transistor is 16 V. Because the transistor has a current flow of 4 mA through it and a voltage drop of 16 V across it, the impedance is 4000 Ω (16/0.004 = 4000).

The base current determines the amount of current flow through the collector–emitter of the transistor. The amount of current flow through the collector–emitter determines the impedance of the transistor. The impedance of the transistor changes, but the current determines the *amount* of change.

APPLICATIONS

A good example of how a transistor can be used to control current flow can be seen in the voltage regulator circuit for many automobiles. The amount of voltage an alternator produces is determined by three factors:

1. The number of turns of wire in the stator

2. The strength of the magnetic field of the rotor

3. The speed of rotation of the rotor

Because the number of turns of wire in the stator is fixed and cannot be changed, the voltage cannot be controlled by changing the number of turns of wire in the stator. Likewise, the speed of rotation of the rotor is determined by the speed of the engine. This prevents controlling the output voltage by controlling the speed of the rotor. The strength of the magnetic field of the rotor, however, is determined by the amount of excitation current supplied to the rotor. If the amount of excitation current is controlled, the output voltage can be controlled.

The function of the voltage regulator is to control the output voltage by controlling the amount of excitation current supplied to the field or rotor. The circuit shown in Figure 10-10 is typical of the solid-state voltage regulator used by many automobiles. In this circuit, the DC output of the alternator is connected in parallel with the battery and the automobile circuitry (represented here by a LOAD resistor). The voltage regulator is also connected in parallel with the output of the alternator. Transistor Q_1 is connected in series with the field of the alternator. This transistor is used to control the amount of current flow through the field. Diode D_1 is connected in parallel with the field. The function of diode D_1 is to prevent a high-voltage spike from being produced when the power is turned off and the magnetic field collapses around the windings of the field.

Transistor Q_2 is connected to the base of transistor Q_1. Resistor R_1 is used to limit the current flow to the base of transistor Q_1 and

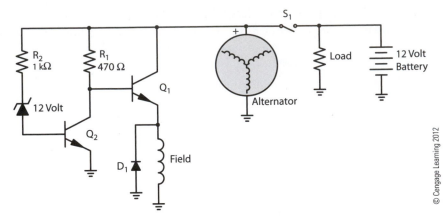

FIGURE 10-10 *Typical voltage regulator circuit for an automobile*

through the collector–emitter circuit of transistor Q_2. Transistor Q_2 is used to "steal" the base current from transistor Q_1. A 12-volt zener diode is connected through a current-limiting resistor, R_2, to the base of transistor Q_2.

The circuit operates as follows:

1. When the engine starts and switch S_1 closes, the 12-volt battery supplies current through resistor R_1 to the base of transistor Q_1. This permits transistor Q_1 to start conducting current through its collector–emitter to the field windings of the alternator. As current flows through the field windings, the magnetic field of the rotor increases, causing an increase in the output voltage of the alternator.

2. When the voltage rises above 12 volts, the 12-volt zener diode begins to conduct. This supplies base current to transistor Q_2.

3. Transistor Q_2 begins conducting part of the base current of transistor Q_1 to ground. This causes transistor Q_1 to conduct less current to the field of the alternator, which reduces the output voltage.

4. If the output voltage should drop too low, the zener diode turns off and stops the supply of base current to transistor Q_2. This permits more base current to flow to transistor Q_1, which causes an increase in the output voltage of the alternator.

UNIT 10 REVIEW QUESTIONS

1. How many layers of semiconductor material are required to make a transistor? _____

_____

2. What are the two major types of transistors? _____

_____

3. Why is silicon used more often than germanium in the production of transistors? _____

_____

4. What is the most common system used in the identification of transistors? _____

_____

5. When an NPN transistor is to be tested with an ohmmeter, does the transistor appear to be two diodes with their anodes connected together or their cathodes connected together? _____

_____

6. When a transistor is in a standard TO5 package, which lead is located closest to the metal tab on the case of the transistor? _____

_____

7. When a transistor is in a standard TO3 case, which lead is connected to the case of the transistor?

_____

8. What controls the amount of current flow through the collector–emitter section of a transistor?

_____

9. What determines the impedance of a transistor? _____

10. When a PNP transistor is to be connected in a circuit, what voltage polarity must be connected to the emitter of the transistor? _____

_____

THE TRANSISTOR SWITCH

OBJECTIVES

After studying this unit, the student should be able to:

- Discuss the use of transistors in a switching application.

- Connect a transistor into a circuit and use it in a switching application.

- Make measurements of transistor voltage drops using test instruments.

The transistor is often used as a simple switch. In this mode, the transistor is used as a digital device in that it has only two states, on or off. An advantage of using the transistor in this mode is its ability to handle power.

SWITCH OPERATION

Assume a circuit where there is 100 volts applied and a maximum current flow of 2 A, Figure 11-1. This circuit will consume 200 watts of power when operating (100 V × 2 A = 200 W). If the transistor in this circuit is in the off state, the voltage drop across the collector and emitter is 100 volts at 0 A. The transistor has to dissipate 0 watt of power (100 × 0 = 0). If the transistor is turned completely on and the base driven into saturation, the voltage drop between the collector and the emitter will be about 0.3 volt. The transistor must dissipate 0.6 watt (0.3 × 2 = 0.6). Notice that a transistor used in this manner has the ability to control a large amount of power.

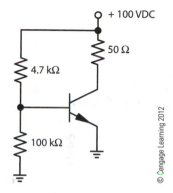

FIGURE 11-1 *Basic transistor circuit*

A transistor is generally considered to be turned completely on when the voltage drop between the collector and the emitter is about 0.7 volt. If the transistor is driven into **saturation**, the voltage drop between the collector and the emitter will be about 0.3 volt. (Note: A transistor is driven into saturation by furnishing more current to the base of the transistor than is needed for normal operation.) If a transistor requires 10 mA of base current to turn it on, 10 mA is considered to be the maximum base current necessary to turn the transistor completely on so that the voltage drop between the collector and the emitter is about 0.7 volt. If the base of this same transistor was furnished with 15 mA, the transistor would go into saturation. The voltage drop between the collector and the emitter will be about 0.3 volt.

Voltage Drop

There is reason for concern about a difference in voltage drop of 0.4 volt (0.7 − 0.3 = 0.4). Assume a transistor is:

- connected into a circuit that has a current of 20 A,
- has a voltage drop of 0.7 volt, and
- will have to dissipate 14 watts of heat (0.7 × 20 = 14).

    If the same transistor:

- is driven into saturation so that the voltage drop between the collector and the emitter is only 0.3 volt,
- it will have to dissipate only 6 watts of heat (0.3 × 20 = 6).

There is a significant difference in the amount of heat the transistor must dissipate.

Saturation: When a transistor is driven into saturation, its frequency response is reduced. It cannot recover or turn off as quickly as it can under normal conditions. For instance, if a transistor is rated to operate at 1 megahertz (MHz), it may not be able to operate above 0.5 MHz when driven into saturation. This reduction of a transistor's ability to operate at a high frequency is not generally a problem in industrial circuits, however. Most motor controllers and induction heating equipment seldom operate above about 1000 Hz.

Transistor or Mechanical Switch?

A transistor is used as a switch in place of a regular mechanical switch because it can operate several thousand times a second. It will last longer than a regular mechanical switch even at that switching rate. The versatility of the transistor switch permits it to be operated by a variety of sensing devices such as devices that sense light, sound, temperature, and magnetic induction.

APPLICATIONS

A very common application for this switch is the electronic ignition system of most automobiles. The collector-emitter section of a transistor is connected in series with the primary of the ignition coil. A small induction coil located in the distributor is connected to the base of the transistor. When a magnet is moved past the induction coil, a voltage is induced into the coil. This voltage is used to trigger the base of the transistor that operates the ignition coil, Figure 11-2.

FIGURE 11-2 *Transistor switch operated by magnetic induction*

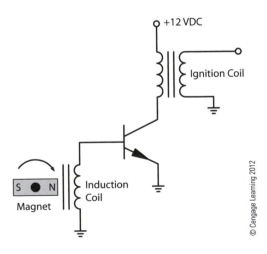

© Cengage Learning 2012

Another common use of the transistor switch is the optoisolator found in industry. A light-emitting diode is used to trigger a photo-transistor. The LED is controlled by the brains of the circuit, and the transistor is used to operate the power handling part of the circuit, Figure 11-3. Optoisolation is used to keep voltage spikes, caused by switching large currents on and off, from "talking" to the electronic control section of the circuit.

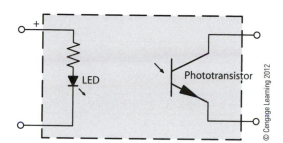

FIGURE 11-3 *Optoisolator*

UNIT 11 REVIEW QUESTIONS

1. When a silicon transistor is considered to be completely turned on, how much voltage is dropped across the collector-emitter section? _____
_____

2. When a transistor is driven into saturation, how much voltage is dropped across the collector-emitter section? _____
_____

3. How is a transistor driven into saturation? _____
_____

4. Why is it desirable to drive a transistor into saturation when it is used for switching? _____
_____

5. What is the disadvantage of driving a transistor into saturation? _____
_____

UNIT 12

THE TRANSISTOR AMPLIFIER

OBJECTIVES

After studying this unit, the student should be able to:

- Discuss amplification of electronic signals.
- Bias a transistor for use as an amplifier.
- Construct a transistor amplifier from discrete components.

Transistor amplification is a subject that seems, at first, to be difficult to understand. Amplification occurs when a small current through the base and emitter is acted upon by the transistor. This results in a larger current moving through the collector to the emitters. The transistor amplifies a signal produced by the base current.

PRACTICAL APPLICATION

Imagine a hydraulically controlled water valve, Figure 12-1. The valve is connected into a water system that has a flow rate of 100 gallons per minute (gpm) and a pressure of 100 pounds per square inch (psi). The valve is controlled by a separate water system that has a flow rate of 1 gpm and a pressure of 1 psi. When 1 gpm of water flows in the control line, the signal causes the main valve to open completely and one hundred gallons of water to flow per minute through the main valve. When the main valve is closed, the pressure across the valve is 100 psi. When the valve is completely open, the pressure is 0 psi. A flow of 1 gpm in the control line causes a flow of 100 gpm through the main valve. This is a hydraulic amplifier that has a ratio of 100:1.

The water flowing in the control line is not increased by the main valve. It is the signal from the control line that causes an increase of water flow.

Operation as an Amplifier

The transistor operates in a similar manner. The transistor shown in Figure 12-2 has a current flow of 100 mA through the collector-emitter and a current of 1 mA through the base. The voltage drop across the transistor will be about 0.7 volt. If the base current is 0 A, the collector-emitter will have a current flow of 0 A and the transistor will have a voltage drop of 20 volts. The base current itself is

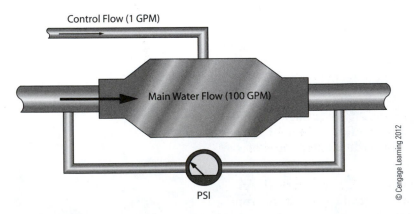

FIGURE 12-1 *Hydraulic amplifier*

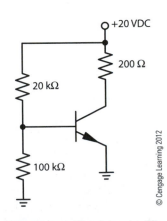

FIGURE 12-2 *Transistor amplifier*

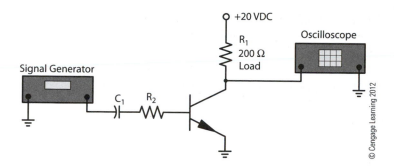

© Cengage Learning 2012

FIGURE 12-3 *Unbiased transistor amplifier*

not increased by the collector-emitter. The 1 mA signal of the base causes 100 mA of collector-emitter current to flow. Therefore, the increase of current flow in the circuit is caused by the 1 mA signal of the base.

When a transistor is to be used as an amplifier, it must be **biased**. Bias means to present or precondition. For example, first consider a transistor amplifier that has not been biased, Figure 12-3.

Circuit Operation

In this circuit, a signal generator supplies an AC signal with a peak voltage of 5 volts. It is assumed that a base-emitter current of 1 mA is required to turn the transistor completely on. Resistor R_1 is used to limit the current flow through the collector-emitter circuit and is the load resistor. Resistor R_2 limits the amount of base-emitter current, and capacitor C_1 blocks any DC component from the signal generator. This ensures that only an AC signal will be applied to the base of the transistor. An oscilloscope is connected across the collector-emitter of the transistor. When the transistor is turned completely off, there will be a voltage drop of 20 volts across the collector-emitter. When 1 mA of base-emitter current flows, the transistor will be turned completely on and a voltage drop of about 0.7 volt will appear across the collector-emitter. If the transistor is a linear device, an increase of base-emitter current will produce a proportional increase in collector-emitter current. When the base current is 0, there is no current flow through the collector-emitter and the transistor has a voltage drop of 20 volts.

1. Now assume that a current of 0.25 mA flows through the base-emitter circuit. This produces a current flow of 24 mA through the collector-emitter of the transistor and a voltage drop of 15.2 volts.

2. If the base current is increased to 0.5 mA, a current of 48 mA flows through the collector-emitter circuit and a voltage drop of 9.7 volts appears across the transistor.

FIGURE 12-4 *AC input signal*

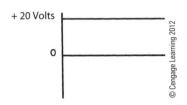

FIGURE 12-5 *20 volts are dropped across the transistor.*

3. If the base-emitter current is increased to 1 mA, the transistor is turned completely on, a current of 98 mA flows through the collector-emitter circuit, and the transistor has a voltage drop of 0.7 volt.

The signal generator is producing an AC signal, as shown in Figure 12-4. When the base current is 0, a voltage of 20 volts will be dropped across the collector-emitter, as shown in Figure 12-5. As the base-emitter current begins to rise in the positive direction, the impedance of the collector-emitter of the transistor decreases and current begins to flow through the load resistor and transistor. This causes the voltage across the collector-emitter to decrease also. This decrease continues until the sine wave current applied to the base reaches its peak value and begins to fall back toward 0. As the base-emitter current begins to fall back toward 0, the collector-emitter current decreases and the voltage across the transistor increases. The output waveform of the transistor is inverse, or opposite, the waveform applied to the base. Notice that as the base current increases, the voltage across the collector-emitter decreases.

When the AC current applied to the base-emitter reaches 0, the transistor is completely turned off and the voltage drop across the collector-emitter is 20 volts. As the base current continues in the negative direction, the transistor cannot turn off more than it is when the base current is 0, so the voltage waveform is not repeated at the output of the transistor. This causes one-half of the waveform to be cut off, as shown in Figure 12-6.

Biasing the Transistor

In order for the transistor to be able to reproduce the base signal, it must be biased. This is done by furnishing a DC current to the base of the transistor and adjusting the current to a point that the transistor is turned half on, Figure 12-7. When resistor R_3 has been adjusted properly, the voltage drop across the collector-emitter will be about 10 volts. This will permit the transistor to be able to turn on as well as turn off. If the base current increases in the positive direction, the transistor will begin to turn on and voltage drop across the collector-emitter will become less than 10 volts. This will continue to happen

FIGURE 12-6 *Output signal of unbiased amplifier*

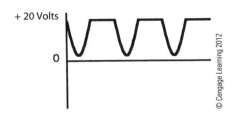

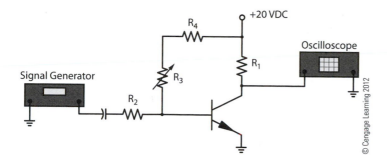

FIGURE 12-7 *Biasing a transistor*

until the base-emitter current reaches its peak value and begins to turn off. When the base current reaches 0, the voltage across the collector-emitter is back to 10 volts. As the base-emitter current becomes negative, the transistor turns off more and the voltage across the collector-emitter becomes greater than 10 volts.

When the voltage applied to the base of the transistor becomes positive, it aids the positive voltage already being applied to the base by resistors R_3 and R_4. This causes the transistor to turn on more and drop less voltage across the collector-emitter. When the voltage applied to the base becomes negative, it is in opposition to the bias voltage and causes the transistor to turn off and produce a greater voltage drop across the collector-emitter. The voltage waveform applied to the base is shown in Figure 12-8, and the voltage waveform produced across the collector-emitter is shown in Figure 12-9. The voltage waveform applied to the base has now been reproduced across the collector-emitter although it is inverted.

FIGURE 12-8 *AC input signal*

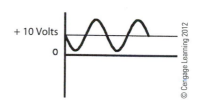

FIGURE 12-9 *Output signal after biasing.*

APPLICATIONS

A good example of how a simple transistor amplifier may be used in an industrial application can be seen in Figure 12-10. In this example, a relay is to be controlled by a light beam. A photodiode, D_2, is used to sense the presence of the light beam. A photodiode is a device that will not conduct when it is in darkness but will conduct when in the presence of light. The photodiode, however, cannot control the amount of current needed to operate the coil of the relay. For this reason, the collector of transistor Q_1 has been connected to the relay coil, and the emitter has been connected to ground. Resistor R_2 is used to ensure the transistor will remain turned off when the photodiode is in darkness. Resistor R_1 limits the amount of current flow through diode D_2 to the base of the transistor when the photodiode turns on. Diode D_1 is used to prevent a spike voltage from being produced when transistor Q_1 turns off and the magnetic field around the relay coil collapses. In this circuit, only a few milliamperes

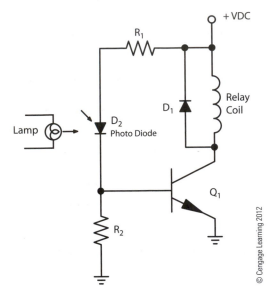

FIGURE 12-10 *Presence of light permits relay to turn on.*

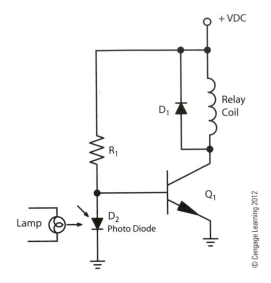

FIGURE 12-11 *The presence of light causes the relay to turn off.*

of base-emitter current will flow when the lamp turns on and causes the photodiode to conduct. These few milliamperes are used to control several hundred milliamperes of current through the collector-emitter needed to operate the coil of the relay.

Figure 12-11 illustrates a variation of the circuit shown in Figure 12-10. In this circuit, the relay will turn on when the lamp is turned off. When the lamp is turned on, the relay will turn off. When the lamp is turned off, the photodiode has a very high impedance. This permits almost all of the current through resistor R_1 to flow through the base-emitter of the transistor. This permits the transistor to turn on and allow current to flow through the coil of the relay. When the lamp turns on, the photodiode exhibits a very low impedance. This causes almost all the current flowing through resistor R_1 to flow to ground instead of through the base-emitter of the transistor. The transistor turns off and stops the flow of current through the relay coil.

UNIT 12 REVIEW QUESTIONS

1. What does the word *bias* mean? _____

2. When a transistor is to be used as an amplifier, it must be biased. Explain how to bias the transistor.
_____
_____

3. Why is it necessary to bias a transistor when it is to be used as an amplifier? _____
_____
_____

4. A transistor is connected into a circuit as an amplifier. The power supply furnishes 12 volts to the transistor. When biasing the transistor, how much voltage should be dropped across the collector-emitter section of the transistor? _____
_____

5. A transistor is connected to operate as an amplifier. When a voltage with a peak-to-peak value of 0.5 volt is applied to the base, it produces a signal of 25 volts peak to peak at the output. What is the gain of this amplifier? _____
_____

UNIT 13

THE DARLINGTON AMPLIFIER

OBJECTIVES

After studying this unit, the student should be able to:

- Discuss the operation of a Darlington amplifier.

- Compute the gain of a Darlington amplifier circuit.

- Construct a Darlington amplifier using discrete components.

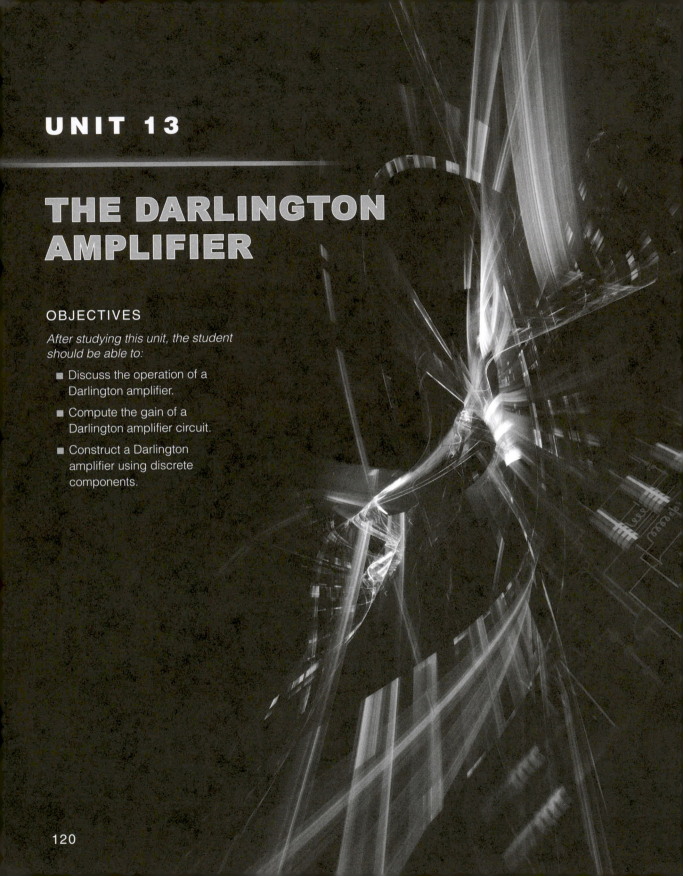

The Darlington amplifier circuit is one of the most used circuits in industry. It is easily made by connecting the emitter of one transistor to the base of another transistor, Figure 13-1.

The Darlington amplifier circuit can have a gain of several thousand. **Gain** is a comparison of the base current to the collector–emitter current. For instance, if 0.1 A of base current is needed to control 1 A of collector–emitter current, the transistor has a gain of 10 (1/0.1 = 10). If 0.01 A of base current is needed to produce 1 A of collector–emitter current, the transistor has a gain of 100 (1/0.01 = 100). Gain can vary from one type of transistor to another. It is measured in a term called *beta*. Transistors have a listed beta that is generally given in a transistor reference book. The beta of a transistor is found by dividing the collector–emitter current by the base current. Beta can apply to a circuit as well. If 1 A of circuit current is being controlled by 1 mA, the circuit gain is 1000 (1/0.001 = 1000).

The Darlington amplifier is used when a high gain is required. Assume that:

- A circuit has a current flow of 10 A and is to be switched on and off by a power transistor.

- The power transistor requires a base current of 10 mA to saturate the transistor.

- The power transistor is to be controlled by a photocell that has a current output of 100 microamperes (µA), Figure 13-2.

The sensor device, which is the photocell, cannot supply nearly enough base current to control the power transistor. The photocell can supply only 1/100 the amount of base current needed to drive the transistor into saturation (10 mA/0.1 mA = 100). However, if a Darlington amplifier circuit is used, control is easily accomplished, Figure 13-3. The photocell is now used to operate a small driver transistor, which in turn supplies the base current to the power transistor.

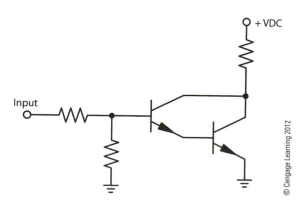

+VDC

Input

© Cengage Learning 2012

FIGURE 13-1 *Darlington amplifier*

FIGURE 13-2 *A photocell supplies transistor base current*

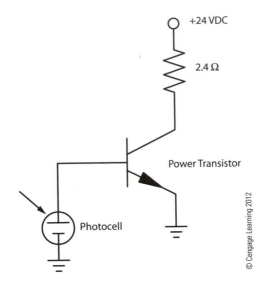

FIGURE 13-3 *Darlington amplifier controlled by a photocell*

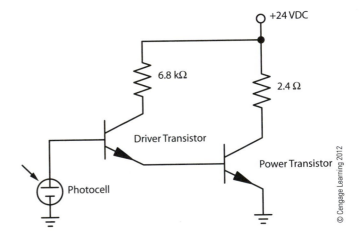

Another way in which the Darlington amplifier is used in industry can be seen in the example circuit in Figure 13-4. In this circuit, transistors are used to control a large amount of current to a load. Three transistors, Q_1, Q_2, and Q_3, are connected in parallel to permit them to share the current necessary to operate the load. Assume that each transistor has a maximum voltage rating of 200 volts and a maximum current rating of 20 A. This circuit would be able to control a voltage of 200 volts and 60 A to the load.

When transistors are used in this manner, they must be driven into saturation to produce as small a voltage drop across the collector–emitter as possible. Recall that a transistor is driven into

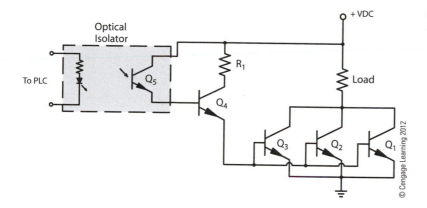

FIGURE 13-4 *Double Darlington amplifier circuit*

saturation by supplying it with more base-emitter current than is normally necessary to turn it completely on.

In this circuit, a programmable logic controller is used to turn the circuit on or off. The programmable logic controller uses optical isolation to prevent voltage spikes and electrical noise from interfering with the central processor unit. Because the output current of the programmable logic controller is much less than that required to drive the three power transistors into saturation, another transistor is used to provide base current to the three transistors controlling current flow to the load. This circuit actually contains a double Darlington amplifier. Transistor Q_4 is used as the driver transistor to supply base current to the three transistors controlling current to the load. The photo transistor, Q_5, is the output of the programmable logic controller. Transistor Q_5 is used as a Darlington drive for transistor Q_4.

A circuit of this type can have a gain of more than a hundred thousand. In this example, the programmable logic controller furnishes a few milliamperes to the light-emitting diode, which in turn can control the amount of current needed to operate the load.

UNIT 13 REVIEW QUESTIONS

1. How is a Darlington amplifier constructed? _____
_____

2. What unit is used to measure the gain of a transistor or circuit? _____
_____

3. In a Darlington amplifier circuit, a base current of 15 microamperes (μA) causes a current of
100 milliamperes (mA) to flow through the circuit. What is the gain of this circuit? _____
_____

4. In a Darlington amplifier circuit, what is the "driver transistor"? _____
_____

5. A Darlington amplifier circuit has a beta of 5000. The power transistor must have a collector–
emitter current of 1.5 A. What is the minimum amount of base current that can be applied to the
driver transistor? _____
_____

UNIT 14

FIELD EFFECT TRANSISTORS

OBJECTIVES

After studying this unit, the student should be able to:

- Discuss different types of field effect transistors.

- Discuss the differences between DE-MOSFETs and E-MOSFETs and JFETs.

- Compare the operation of junction transistors and field effect transistors.

- Connect a field effect transistor circuit.

There are some significant differences between field effect transistors and junction transistors. The field effect transistor has several definite advantages over the junction transistor in various areas. One advantage is that field effect transistors are unipolar rather than bipolar. Junction transistors depend on the movement of both electrons and holes. Field effect transistors depend on only one of the charge carriers, which permits them to operate at higher speeds. FETs exhibit faster switching speeds and higher cutoff frequencies. Other advantages of the FET are:

- They are voltage operated devices rather than current operated. They more closely exhibit the characteristics of vacuum tubes than they do junction transistors.

- In the off state, they exhibit very high input impedance.

- The current flow through the FET is virtually constant with respect to voltage at a specific bias level.

- The current change is inversely proportional rather than directly proportional to temperature.

Field effect transistors (FETs) are so named because they control the flow of current through them with an electric field. There are two basic types of FETs: the junction field effect transistor (JFET) and the metal oxide semiconductor field effect transistor (MOSFET). MOSFETs are often referred to as insulated gate field effect transistors (IGFET). A MOSFET and an IGFET are the same. The first one to be discussed will be the JFET.

JUNCTION FIELD EFFECT TRANSISTORS

JFETs can be divided into two basic types, the N-channel and the p-channel. The schematic symbols for both are shown in Figure 14-1. The pins of both the N-channel and P-channel JFETs are labeled *drain*, *source*, and *gate*. The difference in the two types of JFETs is the polarity of voltage they are connected to. When connecting an n-channel device, the drain connects to the more positive voltage and the source and gate connect to a more negative voltage. When

FIGURE 14-1 *Junction field effect transistor symbols*

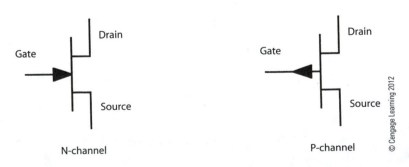

N-channel   P-channel

© Cengage Learning 2012

the N-channel device is used, the drain connects to a more negative voltage, and the source and gate connect to a more positive voltage.

JFET OPERATION

A junction field effect transistor is constructed from a piece of N- or P-type material with an electrical field depletion region inserted in it. Figure 14-2 illustrates the construction of an N-channel JFET. The channel itself is constructed from a piece of N-type material. Two sections of N-type material form the field depletion region. Both of these two sections are connected to the gate pin, although the connection is not shown in the diagram. If a source of voltage is connected to the source and drain, and no source of voltage is connected to the gate, electrons are free to flow through the channel as shown in Figure 14-3. The gate of JFET controls the current flow through the channel. If the gate of an N-channel JFET is connected to a more positive source of voltage than that connected to the source pin as shown in Figure 14-4, the electrical field depletion region becomes smaller and more current is permitted to flow. This, however, is not the manner in which the gate of a JFET is connected. If the JFET were to be connected in this manner, a forward PN junction would be formed and the device could be damaged. Figure 14-5 illustrates what happens when the gate is connected to a voltage source more negative than the voltage connected to the source pin. If the negative voltage connected to the gate becomes high enough, the flow of current through the channel will be stopped completely by the expanding electrical field region. The electrical field region is used to "pinch off" the electron flow through the channel. Notice that

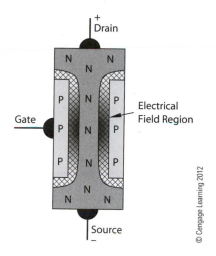

FIGURE 14-2 *Construction of N-channel JET*

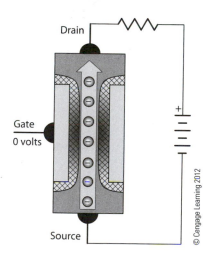

FIGURE 14-3 *Electrons are free to move through the channel.*

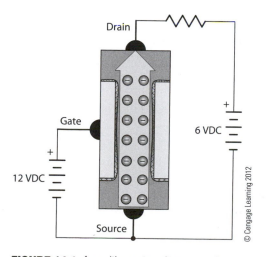

FIGURE 14-4 *A positive gate voltage permits more current to flow.*

FIGURE 14-5 *Enough negative gate voltage can stop the flow of current.*

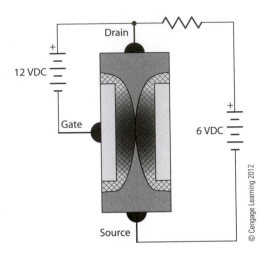

the amount of current flow through the channel is dependent on the amount of voltage applied to the gate, not on the amount of *current* flow through a base. For this reason, the JFET is said to be a voltage-operated device. In fact, junction field effect transistors operate almost identically to old-style vacuum tubes.

JFET IMPEDANCE

One of the great advantages of any field effect transistor is the impedance of the gate circuit. The JFET is normally operated with the voltage connected to the source and gate junction reverse biased. This means that the impedance of the gate circuit is that of a reverse biased diode. This impedance can vary from one device to another, but it is generally in the range of about 20,000 MΩ. This means that a voltage of 20 volts between the gate and source will result in a current flow of about one nanoampere (1×10^{-9}). For this reason, field effect transistors are generally considered to require no gate current when operating.

MOSFETS

MOSFETs differ from JFETs in that there is no PN junction structure. The gate of the MOSFET is insulated from the channel by a layer of silicon dioxide (SiO_2). There are two types of MOSFETs, the depletion-enhancement MOSFET (DE-MOSFET) and the enhancement-only MOSFET (E-MOSFET). Both of these types can be N-channel or P-channel. Because there is an insulator separating the gate from the channel, the input resistance of a MOS-FET is much greater than that of a JFET. This is generally about 100 times greater, which gives the MOSFET an input resistance of about two billion ohms. Typical source to gate current is in the region of 10 pico-amperes (10×10^{-12}) with a voltage of 20 volts. The basic structure arrangement for an N-channel DE-MOSFET is shown in Figure 14-6. The following is an explanation of the operation of an N-channel device. A P-channel device would operate in the same manner with opposite polarity. DE-MOSFETs operate by depletion or enhancement of the channel current carriers as opposed to electrical field depletion. If there is no voltage applied to the gate pin, current is free to flow through the channel as it is in the JFET. When an N-channel DE-MOSFET is operated in the depletion mode, a voltage more negative than that connected to the source is applied to the gate. The negative voltage applied to the gate repels conducting electrons in the channel and leaves positive ions in their place. The electrons are repelled into the P-type substrate region of the device. This *depletion* of electrons reduces the conductivity of the channel, causing less current to flow through the channel. The greater the negative voltage applied to the gate, the greater the depletion of N-channel electrons. If the negative voltage becomes high enough, the N-channel electrons become totally depleted and current flow through the channel stops.

If the DE-MOSFET is operated in the enhancement mode, a voltage more positive than that applied to the source is applied to the gate. This positive voltage attracts electrons from the P-type substrate region into the N-channel region. This increases or *enhances* the conductivity of the N-channel region, which permits more current to flow. Because the gate of the MOSFET is separated from the channel by an insulator and is not a PN junction, it can be operated with either positive or negative voltage connected to the gate.

The schematic symbols for both N-channel and P-channel DE-MOSFETs are shown in Figure 14-7. Notice the gate is shown not connected to the channel. The symbol also shows the substrate as being connected to the source. This is generally, but not always, true.

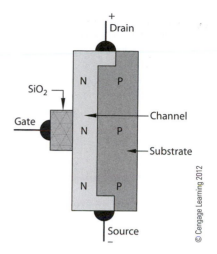

FIGURE 14-6 *N-channel DE-MOSFET*

© Cengage Learning 2012

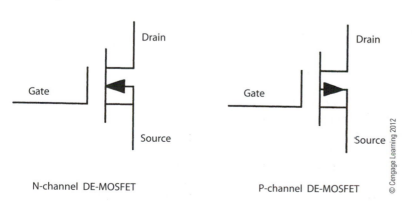

N-channel DE-MOSFET   P-channel DE-MOSFET

FIGURE 14-7 *Schematic symbols of DE-MOSFETs*

© Cengage Learning 2012

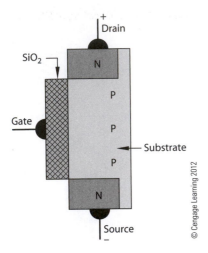

FIGURE 14-8 *Structure of an n-channel E-MOSFET*

E-MOSFETS

E-MOSFETs are structured differently from DE-MOSFETs. The structure of an E-MOSFET is shown in Figure 14-8. Notice that the E-MOSFET does not contain an actual channel as the DE-MOSFET does. The N-type substrate extends to the silicon dioxide insulation layer. If a voltage is connected to the drain and source and no voltage is connected to the gate, there will be no current flow through the device. In order for current to flow through the device, a positive voltage must be applied to the gate. This causes electrons to be attracted from the N-type substrate and form a conducting channel beside the insulator, as shown in Figure 14-9. The higher the voltage applied to the gate, the more conductive the device becomes. The schematic symbols for both N-channel and P-channel E-MOSFETs are shown in Figure 14-10. The schematic symbols use a broken line to indicate there is no physical channel.

TESTING FETS

To test a field effect transistor, it must be known whether it is a JFET or a MOSFET, and whether it is N-channel or p-channel. JFETs can be tested with an ohmmeter. JFETs appear to an ohmmeter to be two diodes connected together with a resistance connected in parallel with them. The polarity of the diodes is determined by whether the device is N-channel or P-channel. Figure 14-11 shows a schematic diagram of what a JFET should look like to an ohmmeter. When checking an N-channel JFET, if the positive lead of the ohmmeter is

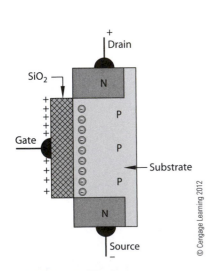

FIGURE 14-9 *Electrons create a channel in the E-MOSFET.*

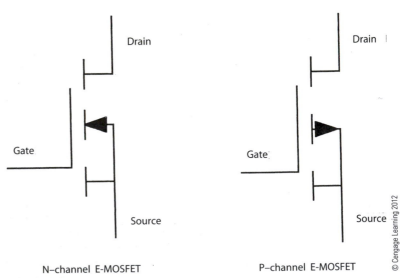

N–channel E-MOSFET   P-channel E-MOSFET

FIGURE 14-10 *Schematic symbols of E-MOSFETs*

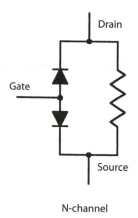

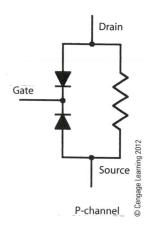

N-channel   P-channel

© Cengage Learning 2012

FIGURE 14-11 *Checking JFETs with an ohmmeter*

connected to the gate, a diode junction should be indicated between gate to source and gate to drain. If the meter leads are reversed, there should be no connection between gate to source or gate to drain. If the ohmmeter leads are connected between the source and drain, some amount of resistance should be indicated. The amount of resistance indicated will depend on the FET. P-channel JFETs can be tested by connecting the negative lead of the ohmmeter to the gate lead and checking for a diode junction between gate to source and gate to drain.

Checking MOSFETs with an ohmmeter is difficult at best. MOSFETs generally come packaged in conducting material that keeps the lead shorted together. This prevents a static charge from building up and damaging the device. MOSFETs can sometimes be damaged just by touching them with a finger. They can be tested with a low-voltage ohmmeter set on its highest resistance scale. If the device is a DE-MOSFET, there should be some continuity between the source and drain, but no continuity between the gate to source or gate to drain. If the device is an E-MOSFET, there should be no continuity indicated between any of the pins.

CONNECTING A FET

As stated previously, the gate of a JFET must be connected to a voltage more negative than the source. Figure 14-12 illustrates the use of a separate power supply connected to the gate of the FET to provide a higher negative voltage than that connected to the source. Most electronic circuits do not employ two separate power supplies, however. To overcome this problem, a resistor is placed in series with the source pin. This forces the source to be at a voltage that is more positive than the gate, Figure 14-13. If the source is at a voltage more positive than the gate, the gate must, therefore, be at

FIGURE 14-12 *The gate must be more negative than the source.*

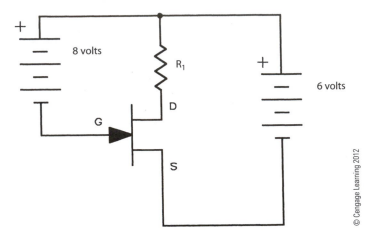

8 volts

R_1

D

G

S

+

6 volts

© Cengage Learning 2012

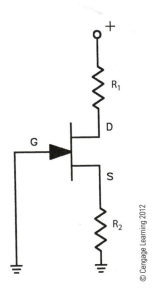

FIGURE 14-13 *The source is more positive than the gate.*

a voltage more negative than the source. The amount of voltage difference between the source and gate is determined by the amount of current flow through the circuit and the resistance of resistor R_2. If a variable resistor is used in this position, the voltage between source and gate can be adjusted. This permits the current flow through the FET to be controlled by adjustment of the variable resistor.

APPLICATIONS

Field effect transistors have found many uses in today's technology. In the circuit shown in Figure 14-14, an FET is used as a simple timer. To understand the operation of the circuit, first analyze what happens when switch S_1 is connected to the TIME position. In this position, the switch is open and the gate of the FET is connected to ground through resistors R_1 and R_2. This permits the transistor to turn on and the LED glows. When the switch is moved to the RESET position, the gate of the FET is connected to a positive voltage, which turns the transistor off. Capacitor C_1 is charged to the value of the applied voltage. When the switch is moved back to the TIME position, the capacitor begins to discharge through resistors R_1 and R_2. When the voltage across the capacitor has become low enough, the FET begins to turn on again. Because the transistor turns on at a gradual rate, the LED will begin to glow faintly and then brighten as the capacitor discharges and permits the FET to turn on more.

The amount of delay time is determined by the size of capacitor C_1 and the sum of resistors R_1 and R_2. Resistor R_2 is used to limit current in the event resistor R_1 should be set for zero ohm. Resistor R_3 limits current flow through the FET and the LED.

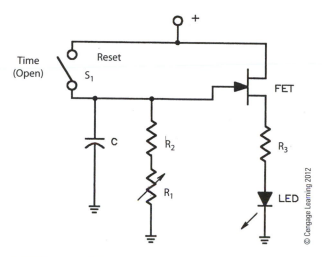

FIGURE 14-14 *FET timer circuit*

FIGURE 14-15 *Auto-ranging digital multimeter*

One of the most common applications for field effect transistors are digital multimeters, Figure 14-15. The circuit shown in Figure 14-16 is the basic schematic of a multirange voltmeter. The input impedance of the meter will be 10 MΩ regardless of the setting of the voltage range switch. The FETs have taken the place of vacuum tubes, which were used to operate high-impedance meters for many years.

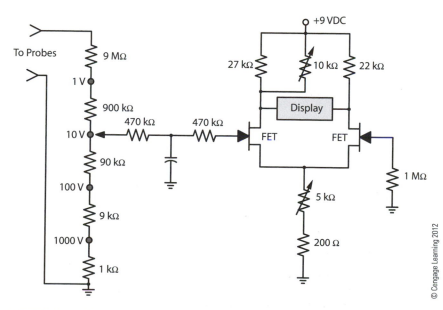

FIGURE 14-16 *Multirange digital voltmeter*

UNIT 14 REVIEW QUESTIONS

1. What is the typical gate resistance of a JFET? _____

_____

2. What does JFET stand for? _____

_____

3. When using a p-channel JFET, must the voltage applied to the gate be more positive or more
negative than the source to turn the device off? _____

_____

4. What are the two types of MOSFET? _____

_____

5. Is an IGFET the same as a JFET or a MOSFET? _____

_____

6. Which type of FET does not contain a channel of p- or n-type material? _____

_____

7. Which type of FET operates by using electrical field depletion? _____

_____

UNIT 15

CURRENT GENERATORS

OBJECTIVES

After studying this unit, the student should be able to:

- Discuss the operation of a current generator.

- List common uses for current generator circuits.

- Construct a constant current generator using a field effect transistor.

- Construct a constant current generator using a junction transistor.

A current generator circuit is used to produce an output that is sensitive to an amount of current flow as opposed to some amount of voltage. Current generators are found in various applications. One of the more common uses for the current generator is found in digital ohmmeters. Many digital ohmmeters measure the value of a resistor by passing a known amount of current through it and then measuring the voltage drop across the resistor. In Figure 15-1, a current generator is connected to an unknown value of resistance. Now assume that the current generator is producing a constant current of 1 mA (0.001), and that the voltmeter connected across the unknown value of resistance is indicating a voltage of 10 volts. By using Ohm's Law, it can be seen that the resistor has a value of 10,000 Ω.

$$R = \frac{E}{I}$$

$$R = \frac{10}{0.001}$$

$$R = 10,000 \ \Omega$$

Now assume that the value of resistance has been changed and the voltmeter indicates a voltage of 3.6 volts. The resistor has a value of 3600 Ω. It is a simple matter of multiplying any value indicated by the voltmeter by 1000 to indicate the proper amount of resistance.

If the resistance value should become high enough that the circuit cannot furnish the required voltage, the current generator is changed to produce a lower value of current and the multiplication factor is changed. For example, assume the current generator has been changed to produce a constant current of 10 μA (0.000010). If

FIGURE 15-1 . The voltage drop across the resistor is proportional to the amount of current flow.

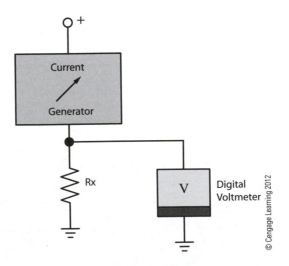

© Cengage Learning 2012

the voltmeter connected across the unknown resistance indicates a value of 2.7 volts, the resistor has a value of 270,000 Ω. The amount of voltage indicated by the voltmeter is now multiplied by a factor of 100,000. Almost any range of resistance can be measured by using the proper amount of current and the correct multiplication factor.

ANALOG SENSORS

Another common use of the current generator is found throughout industry in the use of analog-sensing devices. These devices are used to sense temperature, pressure, humidity, and so on. They are sensors that are designed to operate between some range of settings, such as 50–300°C, or 0–100 psi. These sensors are used to indicate between a range of values instead of just operating in an on or off mode. An analog pressure sensor designed to indicate pressures between 0 and 100 psi would have to indicate when the pressure was 30 psi, or 50 psi, or 80 psi. It would not just indicate whether the pressure had reached 100 psi or not. There are several ways that a pressure sensor of this type can be constructed. One of the most common methods is to let the pressure sensor operate a current generator that produces currents between 4 and 20 mA. It is desirable for the sensor to produce a certain amount of current instead of a certain amount of voltage because it eliminates the problem of voltage drop on lines. For example, assume a pressure sensor is designed to sense pressures between 0 and 100 psi. Also assume that the sensor produces a voltage output of 1 volt when the pressure is 0 psi and a voltage of 5 volts when the pressure is 100 psi. Because this is an analog sensor, when the pressure is 50 psi, the sensor should produce a voltage of 3 volts. This sensor is connected to the analog input of a programmable controller, Figure 15-2. The analog input has a sense resistance of 250 Ω. If the wires between the sensor and the input of the programmable controller are short enough so that there is almost no wire resistance, the circuit will operate without a problem. Because the sense resistor in the input of the programmable controller is the only resistance in the circuit, all of the output voltage of the pressure sensor will appear across it. If the pressure sensor produces a 3-volt output, 3 volts will appear across the sense resistor.

If the pressure sensor is located some distance away from the programmable controller, however, the resistance of the two wires running between the pressure sensor and the sense resistor can cause inaccurate readings. Assume that the pressure sensor is located far enough from the programmable controller so that the two conductors have a total resistance of 50 Ω, Figure 15-3. This means that the total resistance of the circuit is now 300 Ω (250 + 50 = 300). If the pressure sensor produces an output voltage of 3 volts when the pressure reaches 50 psi, there will be a total current flow in the circuit of

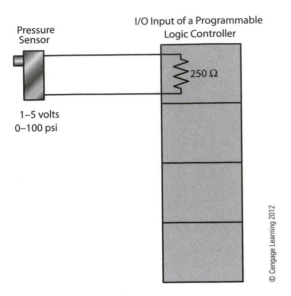

FIGURE 15-2 *A pressure sensor with an output of 1–5 volts*

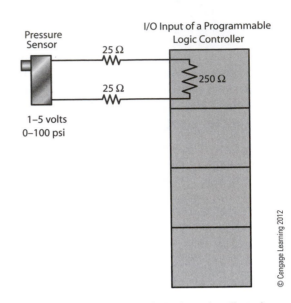

FIGURE 15-3 *Long runs of wire have the effect of adding resistance to the lines.*

10 mA (.010 A) (3/300 = 0.010). Because there is a current flow of 10 mA through the 250 Ω sense resistor, a voltage of 2.5 volts will appear across it. This is substantially less than the 3 volts being produced by the pressure sensor.

If the pressure sensor is designed to operate a current generator with an output of 4–20 mA, the resistance of the wires will not cause an inaccurate reading at the sense resistor. Because the sense resistor and the resistance of the wire between the pressure sensor and the programmable controller form a series circuit, the current must be the same at any point in the circuit. The pressure sensor will produce an output current of 4 mA when the pressure is 0 psi and a current of 20 mA when the pressure is 100 psi. When the pressure is 50 psi, it will produce a current of 12 mA. When a current of 12 mA flows through the 250 Ω sense resistor, a voltage of 3 volts will be dropped across it. Because the pressure sensor produces a certain amount of current instead of a certain amount of voltage with a change in pressure, the amount of wire resistance between the pressure sensor and programmable controller is of no concern.

CONSTRUCTING A CURRENT GENERATOR

A current generator can be constructed using either a field effect transistor or a junction transistor. The field effect transistor circuit will be discussed first. The schematic diagram of a current generator

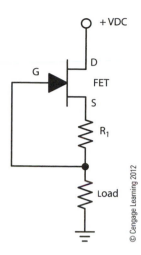

FIGURE 15-4 *A JFET used as a constant current generator*

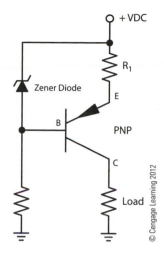

FIGURE 15-5 *A junction transistor used to construct a constant current generator*

using a JFET is shown in Figure 15-4. Resistor R_1 is connected between the source and the gate. It is also connected in series with the load resistor. The value of R_1 determines the amount of current in the circuit. As current flows through the circuit, a voltage drop is produced across resistor R_1. This voltage drop causes the source of the JFET to become more positive than the gate, which forces the gate to begin reducing current flow through the drain-source. If the current flow through resistor R_1 becomes too low, the voltage drop becomes less and the gate permits more current to flow. The amount of load resistance can now be changed and the current will remain constant within the limits of the circuit.

A pnp junction transistor can also be used to construct a current generator. This circuit is shown in Figure 15-5. In this circuit, resistor R_1 is used to determine the amount of current flow through the emitter–collector and load resistor. The zener diode is used to maintain a constant voltage between the base and emitter. In order for current to flow in this circuit, the voltage applied to the emitter must become about 0.7 volt more positive than the voltage applied to the base. Because the zener diode is connected in parallel with resistor R_1, the voltage drop across R_1 must be the same as the zener voltage plus the 0.7 volt needed to operate the transistor. The value of resistor R_1 determines the amount of current that must flow to produce the voltage drop. The value of the load resistor can be changed and the current flow will remain the same as long as the limits of the circuit are not exceeded.

UNIT 15 REVIEW QUESTIONS

1. What is a current generator? _____
_____

2. Explain how resistance is measured using a constant current generator. _____
_____
_____

3. How is the resistance scale changed on an ohmmeter using a constant current generator? _____
_____

4. What is the common current range of analog sensor using current generators? _____
_____

5. What is the advantage of analog sensors using a current generator instead of a constant voltage source? _____
_____

UNIT 16

THE UNIJUNCTION TRANSISTOR

OBJECTIVES

After studying this unit, the student should be able to:

- Discuss the operation of a unijunction transistor.
- Describe the differences between a unijunction transistor and a junction transistor.
- Test a unijunction transistor with an ohmmeter.
- Connect a unijunction transistor in a circuit.

The **unijunction transistor** (UJT) is a special device that has two bases and one emitter. It is a digital device because it has only two states, on or off. It is generally classified with a group of devices known as thyristors, which includes silicon-controlled rectifier (SCR), triac, programmable unijunction transistor (PUT), diac, and the UJT.

CONSTRUCTION OF THE UJT

The unijunction transistor is made by combining three layers of semiconductor material as shown in Figure 16-1. One of the most common UJTs used in industry is the 2N2646. The schematic symbol with polarity connections and the base diagram are shown in Figure 16-2.

Current Flow

The UJT has two paths through it for current flow. One path is from base #2 (B_2) to #1 (B_1). The other is through the emitter and B_1. In its normal state, there is no current flow through the emitter and B_1 path. When the voltage applied to the emitter becomes about 10 volts more positive (higher) than the voltage applied to B_1, the UJT turns on. Increased current flows through the B_1–B_2 path and from the emitter through B_1. Current continues to flow through the UJT until the voltage applied to the emitter drops to a point where it is only about 3 volts higher than the voltage applied to B_1. At this point, the UJT will turn off until the voltage applied to the emitter again becomes about 10 volts higher than the voltage applied to B_1. The unijunction transistor is generally connected into a circuit similar to the one shown in Figure 16-3.

Charge Time

The variable resistor controls the rate of charge time of the capacitor. When the capacitor has been charged to about 10 volts, the UJT turns on and discharges the capacitor through the emitter and B_1. When the capacitor has been discharged to about 3 volts, the UJT turns off and permits the capacitor to begin charging again. By varying the resistance connected in series with the capacitor, the amount of time needed for charging the capacitor can be changed, thereby controlling the pulse rate of the UJT ($T = RC$).

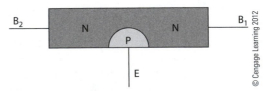

FIGURE 16-1 *Structure of a unijunction transistor*

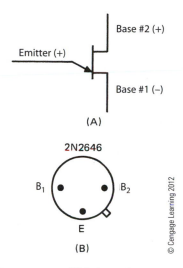

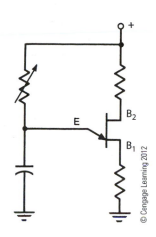

FIGURE 16-2 *(A) Schematic symbol and (B) base diagram*

FIGURE 16-3 *Common unijunction transistor circuit*

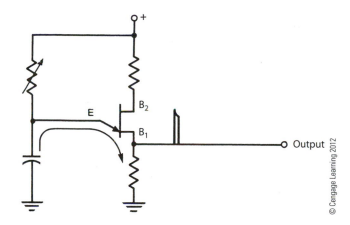

FIGURE 16-4 *Pulse produced by capacitor discharge*

Output Pulse

The unijunction transistor can furnish a large output pulse because the pulse is produced by the discharging capacitor, Figure 16-4. This large pulse is generally used for triggering the gate of an SCR.

The pulse rate is determined by the amount of resistance and capacitance connected to the emitter of the UJT. However, the amount of capacitance that can be connected is limited by the size of the UJT. For instance, the 2N2646 UJT should not have a capacitor larger than 10 microfarads (μF) connected to it. If the capacitor is too large, the UJT cannot handle the current spike produced by the capacitor and can be damaged.

B2
B1
E
© Cengage Learning 2012

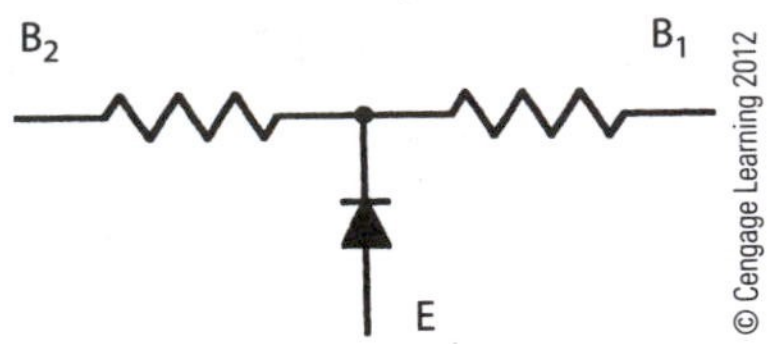

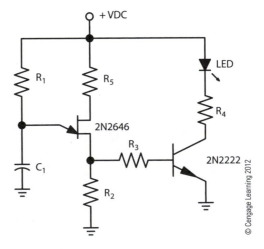

FIGURE 16-5 *Testing a UJT with an ohmmeter*

TESTING THE UJT

The unijunction transistor is tested with an ohmmeter in a manner similar to a common junction transistor. The UJT appears to the ohmmeter as a connection of two resistors connected to a diode, as shown in Figure 16-5. When the positive lead of the ohmmeter is connected to the emitter, a diode junction from the emitter to B_2 and another connection from the emitter to B_1 is seen. If the negative lead of the ohmmeter is connected to the emitter of the UJT, no connection is seen between the emitter and either base.

APPLICATIONS

A circuit using a unijunction transistor is shown in Figure 16-6. This circuit is used as an equipment ON indicator. When some piece of equipment is in operation, a positive voltage is provided at the + input. Capacitor C_1 is charged through resistor R_1. When the voltage of the capacitor exceeds the threshold voltage of the UJT, it turns on and discharges capacitor C_1 to resistors R_2 and R_3. Some of the current flows through resistor R_2 to ground, and some flows through resistor R_3 to the base-emitter of the transistor. This permits the transistor to turn on and provide a current path through the light-emitting diode and resistor R_4. The transistor will remain turned on until capacitor C_1 discharges to a low enough voltage to permit the UJT to turn off. At that point, capacitor C_1 begins charging and the process starts over again. As long as a positive voltage is provided at the + input, the light-emitting diode will flash on and off to indicate that the equipment is in operation.

Programmable Unijunction Transistor (PUT)

The PUT is similar to the unijunction transistor discussed previously except that the PUT permits the operating voltage to be set.

+VDC
LED
R1
R5
R4
2N2646
R3
2N2222
C1
R2
© Cengage Learning 2012

FIGURE 16-6 *Equipment ON indicator*

Recall from the discussion of the unijunction transistor that the breakover or operating voltage was assumed to be 10 volts. The operating voltage of a UJT is set when the device is constructed and cannot be changed.

Another difference between the PUT and the UJT is that the gate of the PUT is connected to the anode of a pn junction, Figure 16-7. The gate of the PUT requires a voltage that is positive with respect to the cathode to set the trigger voltage. The gate of the PUT does not trigger the device, but rather sets the trigger voltage. The circuit shown in Figure 16-8 illustrates the use of a programmable unijunction transistor as an oscillator. Resistors R_2 and R_3 form a voltage divider for the gate. The gate is connected to a voltage that is 8 volts above ground or 8 volts more positive than the cathode. The frequency of the oscillator is determined by the RC time constant set by the values of R_1 and C_1. Capacitor C_1 charges through resistor R_1. When the voltage between the anode and ground reaches approximately 8.6 volts, the PUT turns on and discharges the capacitor. When the voltage drops to a predetermined value the PUT turns off and capacitor C_1 begins charging again.

The programmable unijunction transistor can also be used to phase shift an SCR, Figure 16-9. In this circuit the cathode of the PUT is connected to the gate of the SCR. Resistor R_2 permits the firing rate of the PUT to be adjusted by changing the RC time constant with capacitor C_1.

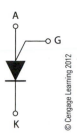

FIGURE 16-7 *Symbol for a programmable unijunction transistor (PUT)*

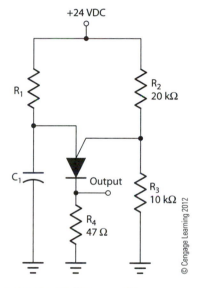

FIGURE 16-8 *Programmable unijunction transistor used as an oscillator*

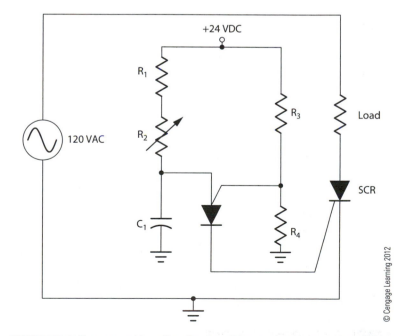

FIGURE 16-9 *Programmable unijunction transistor used to phase shift an SCR*

UNIT 16 REVIEW QUESTIONS

1. The unijunction transistor is a member of what family of electronic devices? _____
_____

2. The connections of a junction transistor are the emitter, base, and collector. What are the
connections of a unijunction transistor? _____
_____

3. In the example explanation in this unit, the voltage applied to the emitter must become 10 volts
more positive than the voltage applied to which base before the transistor will turn on? _____
_____

4. How is the pulse rate of the UJT controlled? _____
_____

5. Why can the UJT furnish a large current spike at the output? _____
_____

6. What is the major function of the UJT? _____
_____

7. What is the main difference between a UJT and a PUT? _____
_____

8. When setting the trigger voltage of a PUT, must the gate voltage be positive or negative with
respect to the cathode voltage? _____
_____

UNIT 17

THE SCR IN A DC CIRCUIT

OBJECTIVES

After studying this unit, the student should be able to:

- Discuss the operation of an SCR when connected to a DC circuit.

- Draw the schematic symbol for an SCR.

- Test an SCR with an ohmmeter.

- Connect an SCR in a DC circuit.

The silicon-controlled rectifier (SCR) is made by bonding four layers of semiconductor material together to form a PN–PN junction, Figure 17-1.

The SCR is called a thyristor because its characteristics are similar to the gas-filled thyratron tube used in industrial electronics for many years. Since the invention of the SCR, other devices with similar operating characteristics have been invented and are also listed as thyristors. Today there is an entire family of devices known as thyristors.

IDENTIFYING THE SCR

SCRs are known as the workhorses of industrial electronics because of their ability to control hundreds of volts and amperes. However, they are also found in smaller sizes with ratings of less than 1 A at only a few volts. SCRs are also found in many different sizes and styles of cases.

Notice in Figure 17-2 that SCRs can be found in many of the same case styles as transistors. Without some means of identifying the component, there is no way of knowing what a device is by looking at the case. SCRs are numbered in a manner similar to transistors. These include:

- A registered 2N number, which can be identified in a semiconductor index

FIGURE 17-1 *Structure of a silicon-controlled rectifier (SCR)*

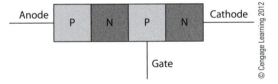

© Cengage Learning 2012

FIGURE 17-2 *Different case sizes and styles for SCRs*

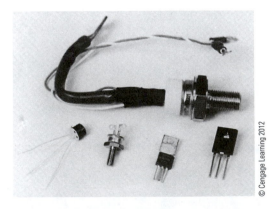

© Cengage Learning 2012

■ A manufacturers' special number, which can be identified with the proper information

■ Numbers by equipment manufacturers, which cannot be identified by anyone other than the manufacturer

OPERATING CHARACTERISTICS

To understand the operating characteristics of the SCR, it is necessary to first understand how it operates when connected in a DC circuit. The schematic symbol for an SCR is shown in Figure 17-3.

The SCR operates very similarly to a relay that is controlled by a set of push buttons and is latched with holding contacts, Figure 17-4. In this circuit, the start button is used to turn relay coil CR on. Once the coil is turned on, the holding contacts, operated by coil CR, close and maintain a current path around the normally open push button, Figure 17-5.

Once the holding contacts have closed, current continues to flow to coil CR even if the start button is released and the push button is again opened. The start button can control when the relay coil is turned on, but once the relay is energized, the start button has no more control over the circuit.

In order to turn the relay off, the normally closed stop button must be pushed to open the circuit to the coil, Figure 17-6. After the relay coil has been de-energized, CR contacts open and the stop button can be released.

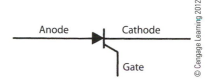

FIGURE 17-3 *Schematic symbol of the silicon-controlled rectifier (SCR)*

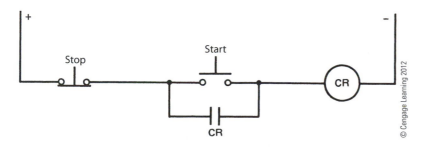

FIGURE 17-4 *Start–stop push-button circuit*

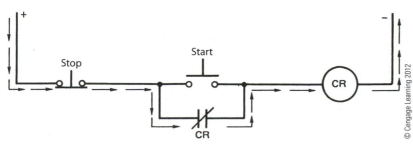

FIGURE 17-5 *Contacts maintain circuit*

FIGURE 17-6 *Current path broken by stop button*

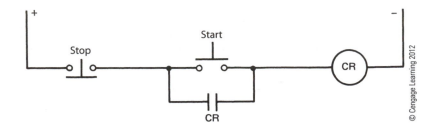

© Cengage Learning 2012

FIGURE 17-7 *An SCR operates similarly to a start–stop push-button circuit.*

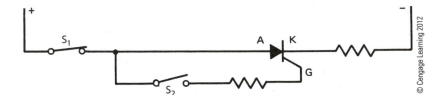

© Cengage Learning 2012

Operation of the SCR

The SCR operates in a manner similar to this relay circuit. The SCR in Figure 17-7 will not conduct until switch S2 is closed and current flows through the gate–cathode circuit. When the gate energizes, the SCR conducts through the anode–cathode circuit. After the SCR is turned on by the gate, switch S2 can be opened (removing current from the gate), and the SCR continues to conduct. When switch S1 is opened, current flow through the anode–cathode circuit is broken, and the SCR turns off. Switch S1 can now be closed and the SCR will remain turned off until the gate is again used to turn it back on. Notice that the gate operates similarly to the start button in the relay circuit. It can turn the SCR on, but not off.

GATE CURRENT: The amount of gate current needed to turn the SCR on varies from one device to another. SCRs designed to control small amounts of power require a gate current of only a few microamperes, whereas SCRs designed to control large amounts of power require a gate current of several hundred milliamperes. However, the amount of gate current needed to fire the SCR is only a small fraction of the amount of current the device is designed to handle through the anode–cathode circuit.

HOLDING CURRENT: The amount of current flowing through the anode–cathode circuit needed to keep the SCR turned on is called the **holding current**. Assume a certain SCR has a holding current of 100 mA. As long as current flowing through the anode–cathode circuit is above 100 mA, the device remains on. If the current drops below 100 mA, it turns off. As with gate current, the amount of

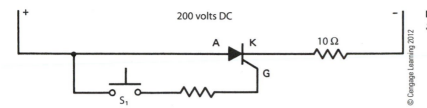

FIGURE 17-8 *The gate will turn the SCR on, but not off.*

current needed to keep an SCR turned on varies from one device to another. A general rule is, the more power an SCR is designed to control, the higher its holding current will be.

Controlling Power

The SCR is the workhorse of industrial electronics because it can control large amounts of power. This is because it has only two states of operation, completely on or completely off.

In the off state of Figure 17-8, the full 200 volts is dropped across the SCR. Because there is no current flowing through the circuit, there is no power being dissipated in the form of heat ($200 \times 0 = 0$). When the gate receives a current pulse through switch S_1, the SCR turns on and about 20 A of current flows through the circuit ($200/10 = 20$). In the on state, the SCR has a voltage drop of about 1 volt. In this circuit, it has to dissipate about 20 watts of heat ($20 \times 1 = 20$); it will have to be heat sinked and thermal compound used.

TESTING THE SCR

SCRs can be tested with an ohmmeter by connecting the positive lead to the anode and the negative lead to the cathode, Figure 17-9. When an ohmmeter is connected across the anode–cathode circuit of the SCR, there is infinite or high resistance. This high resistance value is seen because the SCR is off unless triggered by the gate.

With the leads connected and using a jumper, touch the gate of the SCR to the positive ohmmeter lead. The SCR shows a great decrease of resistance. When the gate is disconnected from the positive lead, the SCR may turn off or continue to conduct, depending on the device. If it has a small holding current, the ohmmeter may supply enough current to keep the SCR turned on. If the ohmmeter cannot supply enough holding current, the SCR will turn off.

It should be noted that many of the high-power SCRs used in industry have an internal resistor connected between the gate and cathode, as shown in Figure 17-10. The purpose of this resistor is to help keep the SCR from false triggering due to line interference. This resistance can be measured with an ohmmeter when testing the

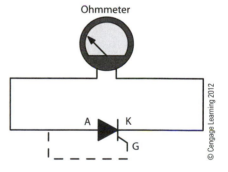

FIGURE 17-9 *Ohmmeter test*

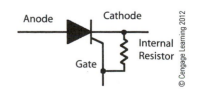

FIGURE 17-10 *SCR with an internal resistor*

SCR and can cause a maintenance electrician who is not aware of the existence of this resistor to diagnose the device as being leaky between the gate and cathode.

APPLICATIONS

A good example of how an SCR can be used in a DC circuit can be seen in Figure 17-11. This is a battery charger circuit that turns the power applied to the battery off when the battery reaches full charge. The transformer is a step-down transformer with a secondary voltage of 28 volts. Notice that the secondary of the transformer has been center tapped. The two diodes, D_1 and D_2, form a full-wave rectifier, with the center tap of the transformer forming the positive output lead. It is assumed that this circuit is to be used to charge a 12 volt lead-acid battery.

When a 12 lead-acid battery reaches full charge, it will exhibit a terminal voltage of 14 volts. If the battery should be charged beyond this point, it will overcharge, which could damage the battery. To prevent overcharging, the battery should be charged to a voltage of less than 14 volts. The ideal voltage to charge a 12 lead-acid battery is generally considered to be 13.8 volts.

In this circuit, the two-diode type of rectifier will produce a peak DC voltage of 19.8 volts ($14 \times 1.414 = 19.8$). When a battery is in a low-charge state, the terminal voltage is low, which permits the power supply to furnish current to the battery. As the state of charge of the battery increases, its terminal voltage will increase to a maximum of 14 volts. The amount of charging current will be much higher during the period of time the battery is in a low state of charge. As the state of charge of the battery increases, the amount of charging current will decrease to a very small amount.

FIGURE 17-11 *Battery charger with automatic turn off*

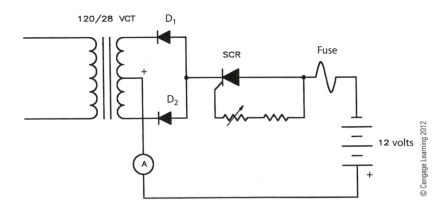

© Cengage Learning 2012

The amount of gate current supplied to the SCR is determined by the amount of resistance connected in the gate circuit, and the difference in voltage between the power supply and the terminal voltage of the battery. When the battery is in a low-charge state, the difference in voltage is great enough to permit enough gate current to flow to turn the SCR on. Once the SCR has been turned on by the gate, it will remain on due to the relatively high-charging current. As the state of charge of the battery increases, two things happen at the same time. The charge current decreases, and the terminal voltage of the battery increases. As the terminal voltage of the battery increases, there is less difference in voltage between the power supply and the battery, which causes the gate current supplied to the SCR to decrease. If the variable resistor connected in the gate circuit of the SCR is adjusted properly, the amount of gate current supplied to the SCR will drop below the level needed to turn it on when the terminal voltage of the battery reaches about 13.8 volts. If the SCR has been selected correctly, the charging current at this point will have dropped below the holding level of the SCR. This permits the SCR to turn off and stop conduction of the charging current completely.

GTO (Gate-Turnoff) SCR

The gate-turnoff SCR is also known by the name gate-controlled switch (GCS). The gate-turnoff SCR operates in a fashion similar to a common SCR in that it is turned on by a gate pulse that is positive with respect to the cathode. Unlike the SCR, however, it can be turned off with a gate pulse that is negative with respect to the cathode. A commonly used schematic symbol for the GTO is shown in Figure 17-12. Although the GTO can be turned off with a negative current pulse on the gate, it requires from 10 to 20 times more negative gate current to turn it off than it does to turn it on.

The greatest advantage of the GTO over a common SCR is its ability to control power in a direct current circuit. Because the anode–cathode current does not have to drop below a holding current level to turn it off, it is much easier to control than an SCR.

FIGURE 17-12 *Schematic symbol for a gate-turnoff SCR*

UNIT 17 REVIEW QUESTIONS

1. How many layers of semiconductor material are used in the production of an SCR?

2. What do the letters SCR stand for? _____

3. Name the three terminals of an SCR. _____

_____

_____

4. When an SCR is connected into a DC circuit, what controls the turn on of current flow through the anode–cathode section of the SCR?

5. How is the current flow through the anode–cathode section turned off after it has been turned on?

_____

6. How many states of operation does the SCR have?

7. What is the advantage of the GTO as compared to the SCR?_____

_____

8. How much more gate current is required to turn a GTO off as compared to the amount of current needed to turn it on?

UNIT 18

THE SCR IN AN AC CIRCUIT

OBJECTIVES

After studying this unit, the student should be able to:

- Discuss the operation of an SCR in an AC circuit.
- Discuss problems of control.
- Connect an SCR in an AC circuit.

When connecting an SCR in an AC circuit, the first thing to remember is that the SCR is a rectifier, and the output of the SCR is pulsating direct current. The only difference between an SCR and a common rectifier diode is that the SCR can be triggered to turn on at a specific point in the waveform.

DC AND AC CONNECTIONS

When the SCR is connected into a DC circuit, the gate can turn the SCR on but cannot turn it off. The current through the anode–cathode circuit of the SCR must drop below the holding current level before the device will turn off.

The same is true for an SCR connected into an AC circuit. The gate still controls the turn on of the SCR. The anode–cathode current must drop below the holding current level for the device to turn off. The main difference in operation is caused by the AC power itself. The gate turns the SCR on, but the AC waveform dropping back to 0 volts will turn it off. It is therefore necessary to retrigger the gate of the SCR for each half cycle of alternating current conducted.

OPERATION OF THE CIRCUIT

Consider the circuit shown in Figure 18-1. In this circuit, the gate of the SCR is connected through a resistor and diode directly to its anode. When the AC voltage applied to the anode rises in the positive direction, current flows through the gate–cathode section of the SCR. When it reaches the trigger point (assume 5 mA for this SCR), the SCR fires and conducts through the anode–cathode section. It conducts as long as the AC voltage remains in the positive direction and the current is above its holding current level. When the AC voltage drops to 0 and begins to increase in the negative direction, the SCR remains turned off. When the voltage applied to the anode again becomes positive, the gate will trigger the SCR on again. The output of SCR is half-wave rectified direct current, Figure 18-2.

Adding the Variable Resistor

Why use an SCR if it does the same thing as a simple junction diode? The SCR can be controlled as to when it turns on, Figure 18-3. A variable resistor has been added to the gate circuit of the figure.

FIGURE 18-1 *Basic SCR circuit*

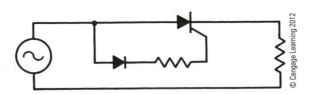

© Cengage Learning 2012

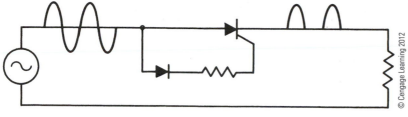

FIGURE 18-2 *SCR with no control of the gate current*

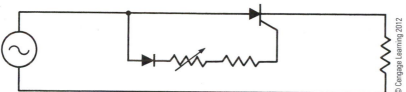

FIGURE 18-3 *Variable resistor controls gate current.*

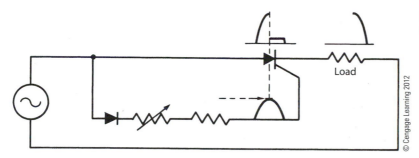

Load

FIGURE 18-4 *SCR fired at the peak of the waveform*

Assume that the gate current must reach a 5 mA level before the SCR will fire. By adjusting the variable resistor, it is possible to determine how much voltage must be applied to the gate before a 5 mA current flows through the gate circuit. By adjusting the resistor to a higher value, it is possible to keep the SCR from firing until the AC voltage has reached its peak value in the positive direction. In this way, the SCR will fire when the AC voltage is at its positive peak. With this setting of the resistor, the SCR will drop half the voltage and the load will drop the other half, Figure 18-4.

REDUCING THE RESISTANCE: By reducing the resistance of the gate circuit, the gate current reaches 5 mA sooner, and the SCR fires earlier in the AC cycle. This causes less voltage to be dropped across the SCR and more to be dropped across the load, Figure 18-5.

If the gate resistance is reduced even more, the 5 mA of gate current will be reached even sooner during the cycle and the SCR will fire earlier. Still less voltage is dropped across the SCR and more voltage is dropped across the load, Figure 18-6.

FIGURE 18-5 *SCR fired before the waveform reaches peak*

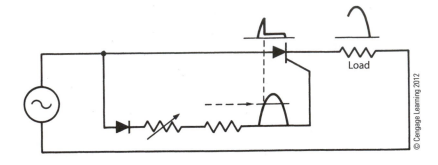

FIGURE 18-6 *SCR fired earlier than in Figure 18-5*

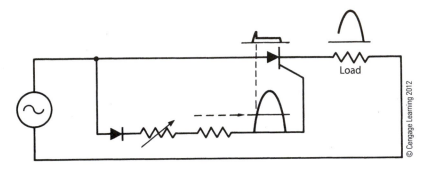

PROBLEMS MET ON THE WAY TO FULL CONTROL

There is a problem with this type of control, however, the SCR only controls half of the positive cycle of alternating current applied to it, Figure 18-5. The latest that the SCR can be fired is when the positive half cycle has reached 90°. This permits the SCR to control only half of the AC positive wave; half voltage is applied to the load when the SCR initially fires. If the load resistor is a light bulb, the bulb will burn at half brightness when it is first turned on. The SCR circuit controls the light bulb from half brightness to full brightness. There are methods of gaining full control of the waveform; these will be discussed in Unit 19.

APPLICATIONS

The circuit shown in Figure 18-7 is an example of how an SCR can be used in an AC circuit. In this circuit, the SCR is used as a touch controller. When the touch plate is touched by a person, the SCR turns on and permits current to flow through the load. The SCR will remain turned on as long as the person continues to touch the plate. This circuit can be used when it is necessary to have a switch with no moving or mechanical contacts.

Notice that a neon lamp has been connected in series with the gate of the SCR. The neon lamp is used because it is a gas-filled tube and depends on ionization of the gas for operation. Most devices

FIGURE 18-7 *Touch switch*

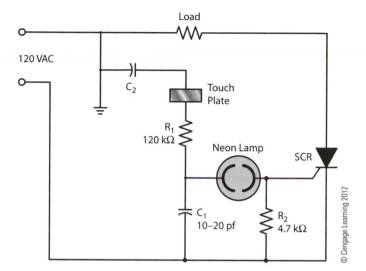

that operate by ionization have a characteristic known as negative resistance. This means that the device will start conduction or turn on at one voltage and continue conduction at a lower voltage. For example, assume that this lamp requires a potential of 15 volts across it to ionize or start conduction. Once the gas in the tube has ionized, it will continue to conduct at a much lower voltage, say 5 volts. As long as the voltage across the tube is greater than 5 volts, it will continue to conduct. When the voltage drops below 5 volts, the tube will stop conduction or turn off, Figure 18-8.

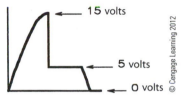

FIGURE 18-8 *The voltage must drop below 5 volts before conduction will stop.*

The circuit operates due to the voltage divider formed by capacitors C_1 and C_2. Capacitor C_2 is actually the body capacitance of a person. The largest amount of voltage will be dropped across the capacitor with the lowest amount of capacitance. When no one is touching the touch plate, the circuit between the touch plate and ground is open. This forms a very low value of capacitance, causing almost all the circuit voltage to be dropped across these two points. When the touch plate is touched, the capacitance of C_2 becomes much greater than that of C_1. This causes the voltage across capacitor C_1 to become higher than 15 volts, causing the neon lamp to ionize and start conduction. When the neon lamp ionizes, its internal impedance drops immediately, permitting capacitor C_1 to discharge and supply a pulse of current to the gate of the SCR. The SCR turns on and allows current to flow through the load.

Resistor R_1 limits current in the gate circuit, and resistor R_2 is used to ensure that the SCR will remain turned off until the neon lamp ionizes and discharges capacitor C_1. The sensitivity of this circuit is determined by the area of the touch plate. If the area becomes great enough, this control will become a proximity detector and will trigger when a person comes near the touch plate without actually touching it.

UNIT 18 REVIEW QUESTIONS

1. When an SCR is connected into an AC circuit, is the output voltage AC or DC? _____

_____

2. After the SCR has been turned on by the gate, what causes the SCR to turn off again? _____

_____

3. Explain the operating difference between an SCR and a junction diode. _____

_____

4. What is the latest point in the positive half cycle that the SCR can be made to fire? _____

_____

5. By reducing the resistance connected in the gate circuit, is it possible to make the SCR fire earlier than 90° or later than 90°? _____

_____

PHASE SHIFTING
AN **SCR**

OBJECTIVES

After studying this unit, the student should be able to:

- Discuss the meaning of phase shifting an SCR.

- Discuss the reasons for phase shifting an SCR.

- Construct a circuit for phase shifting an SCR in an AC circuit.

In order for an SCR to gain complete control of the waveform, it must be **phase shifted**. This means to change or shift the phase of one thing in reference to another. In this case, the concern is with shifting the phase of the voltage applied to the gate with respect to the voltage applied to the anode.

PRACTICAL APPLICATION

Recall that an SCR that has not been phase shifted can control only half of the positive waveform. An SCR that has not been phase shifted has the voltage that is applied to the gate locked in phase with the voltage applied to the anode, Figure 19-1. Assume that the SCR in this example fires when the gate current reaches a 5 mA level.

The variable resistor in Figure 19-2 is adjusted so that the gate current will not reach the 5 mA level until the applied voltage reaches its peak value. The SCR fires when the voltage applied to the anode is at the peak level also. If the variable resistor is decreased in value, the gate current can reach 5 mA sooner, and the SCR will fire sooner during the cycle, Figure 19-3.

FIGURE 19-1 *Voltage applied to the gate in phase with the voltage applied to the anode*

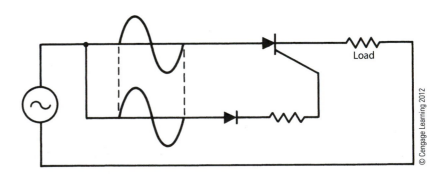

© Cengage Learning 2012

FIGURE 19-2 *SCR fired at the peak of the waveform*

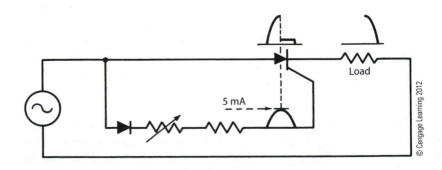

© Cengage Learning 2012

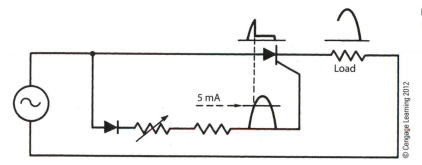

FIGURE 19-3 *SCR fired before the waveform reaches the peak*

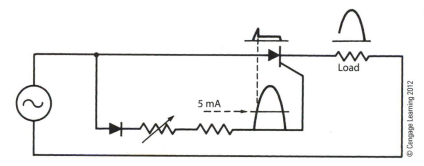

FIGURE 19-4 *SCR fired very early in the waveform*

Each decrease of the variable resistor will cause the gate current to reach 5 mA earlier in the cycle and cause the SCR to fire sooner, Figure 19-4.

Separating Voltage

Notice that the gate controls the firing of the SCR during only half of the positive cycle. As long as the voltage applied to the gate of the SCR is locked in phase with the voltage applied to the anode, the SCR cannot be fired after the AC voltage applied to the anode has reached 90°. To obtain full control of the AC half cycle applied to the anode, the voltage applied to the gate must be out of phase with the voltage applied to the anode. To shift the phase of the voltage applied to the gate, the gate voltage must be separated from the voltage applied to the anode. There are several ways to do this, but for now only one of these methods will be investigated.

CAPACITIVE PHASE SHIFT. One of the simplest ways to phase shift an SCR is to use a small center-tapped transformer to provide trigger circuit isolation from the line, Figure 19-5. The variable

FIGURE 19-5 *Isolation transform used to phase shift SCR*

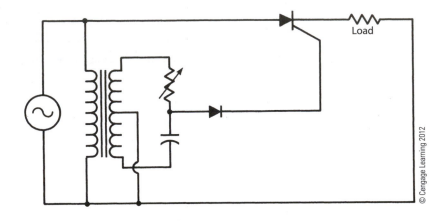

FIGURE 19-6 *SCR fired after the waveform has passed the peak*

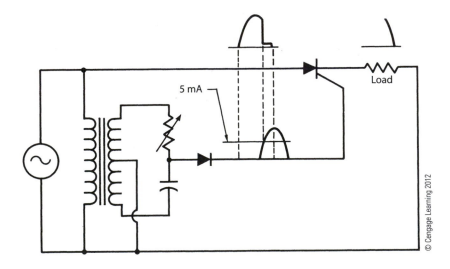

resistor and capacitor in this circuit causes the gate current to be capacitive in comparison to the anode–cathode current. Because the current in a capacitive circuit leads the voltage, the gate current is shifted out of phase with the voltage. The variable resistor determines the amount of phase shifting. By shifting the gate current out of phase with the voltage, the gate current reaches 5 mA after the applied voltage reaches its peak value, Figure 19-6. Because the gate current reaches the 5 mA level after the voltage applied to the anode reaches its peak value, the SCR fires later in the cycle. This causes a higher voltage to be dropped across the SCR and less voltage dropped across the load.

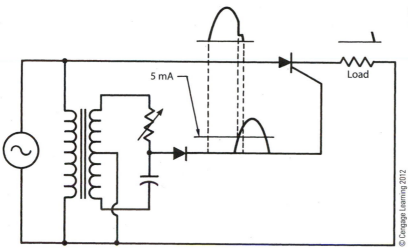

FIGURE 19-7 *SCR fired late in the waveform*

5 mA

Load

If the variable resistor is adjusted so that the gate current leads the voltage by a greater amount, the SCR fires later, Figure 19-7. The SCR therefore drops a greater amount of voltage and the load has less voltage applied to it. By choosing the correct values of resistance and capacitance, the voltage applied to the load can be controlled from 0 to its full value.

UNIT 19 REVIEW QUESTIONS

1. What is phase shifting? _____

_____

2. When phase shifting an SCR, what two things must be phase shifted? _____

_____

3. In the example shown in this unit, what electrical device is used to separate the gate voltage from the anode voltage? _____

_____

4. What electrical device is used to cause the gate current to lead the voltage? _____

_____

5. Why is it necessary to phase shift an SCR when it is connected to an AC circuit? _____

_____

UNIT 20

UJT PHASE SHIFTING
FOR AN SCR

OBJECTIVES

After studying this unit, the student should be able to:

- Discuss the use of a UJT as a phase-shifting device for an SCR.

- Discuss the operation of a Shockley diode.

- Draw the schematic symbol of a Shockley diode.

- Discuss the use of a Shockley diode as a phase-shifting device for an SCR.

- Discuss the characteristics of a silicon unilateral switch.

- Draw the schematic symbol of an SUS.

- Construct a phase-shift circuit for an SCR using a UJT.

The unijunction transistor (UJT) was developed to do the job of phase-shifting SCR circuits. It is an ideal device for this application because its operating characteristic is that of a pulse timer.

An SCR must be phase shifted to gain complete control of the AC waveform supplied to it. The UJT offers an easy way of accomplishing this job, Figure 20-1. This circuit can be broken down into two separate sections. One is the power handling part and consists of the SCR and the load. The other is the brains and consists of the step-down transformer, UJT, and related components.

The transformer in this circuit is used to provide a low voltage to operate the UJT. The UJT is turned off until capacitor C_1 charges to a predetermined voltage. Assume this voltage to be 10 volts. The charge time for the capacitor is determined by its capacitance and the resistance of resistor R_1. When the charge on the capacitor reaches a value of 10 volts, the UJT turns on and discharges the capacitor through resistor R_2. This discharge produces a pulse of current at R_2, which triggers the gate of the SCR. When the voltage across C_1 drops to about 3 volts, the UJT turns off and permits the capacitor to begin charging again. Because the charge time of the capacitor is determined by the resistance of R_1, the pulse rate of the UJT is controlled by varying the resistance of resistor R_1.

The pulses produced by the UJT are entirely independent of the AC voltage connected to the anode of the SCR. Because the UJT can be triggered at any time regardless of the AC waveform, the SCR can be fired at any point during the positive half cycle of alternating current applied to it.

Phase-shifting SCR circuits with the unijunction transistor has become a very common practice in industrial electronic controls. It is important to gain a good understanding of this and other phase-shifting circuits.

FIGURE 20-1 *Unijunction transformer (UJT) phase-shift circuit*

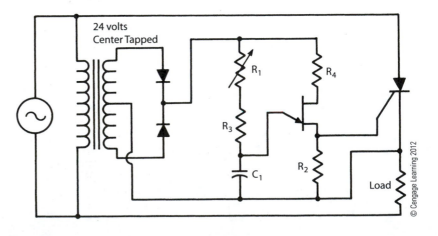

© Cengage Learning 2012

OTHER TRIGGERING DEVICES

Although the unijunction transistor is one of the most common devices used for phase-shifting SCR circuits, it is not the only device that can be used. Two other devices that can be implemented to phase shift an SCR are the four-layer diode, or **Shockley diode**, and the **SUS**, or silicon unilateral switch.

The Shockley diode has a negative resistance characteristic very similar to that of the UJT. The diode will remain turned off until its breakover voltage is reached. At this point, the diode turns on and conducts until the voltage across the device drops to its turn-off value. The breakover, or turn-on, value of voltage is higher than the turn-off value. For example, its turn-on value may be 8 volts and the turn-off value may be 3 volts. The device's ability to conduct current at a lower voltage value than it takes to turn it on makes it useful for phase shifting. The schematic symbol for a Shockley diode is shown in Figure 20-2.

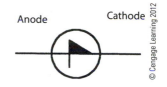

FIGURE 20-2 *Shockley diode*

A simple SCR phase-shifting circuit using the Shockley diode is shown in Figure 20-3. When the AC waveform applied to the anode of diode D_1 becomes positive, the diode turns on and begins to conduct current through resistors R_1 and R_2. This permits capacitor C_1 to begin charging. When the voltage across capacitor C_1 increases to the breakover voltage of the Shockley diode, it turns on and discharges capacitor C_1 through the gate of the SCR. This pulse of gate current causes the SCR to turn on and conduct current through the load. When C_1 has discharged to a voltage that is below the turn-off value of the Shockley diode, it turns off. The next positive half cycle of AC voltage begins charging the capacitor again, and the sequence of events is repeated.

The amount of time necessary for capacitor C_1 to reach the breakover value of voltage for the Shockley diode is determined by the value of capacitance of capacitor C_1 and the sum of the resistances

FIGURE 20-3 *Shockley diode used to phase shift an SCR*

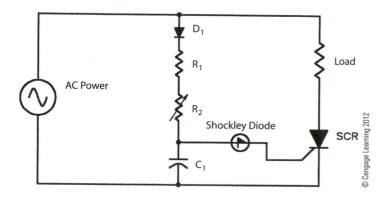

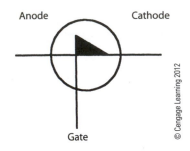

FIGURE 20-4 *Silicon unilateral switch*

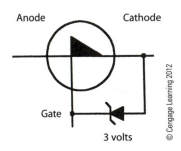

FIGURE 20-5 *Zener diode used to adjust the breakover voltage of an SUS*

of resistors R_1 and R_2. Resistor R_2 has been made variable to permit the charge time of capacitor C_1 to be adjusted. If the values of capacitor C_1 and resistor R_2 are chosen correctly, the SCR can be phase shifted over the entire half cycle.

The silicon unilateral switch is very similar to the Shockley diode. The basic difference between the two devices is that the SUS has a gate lead that permits some control of the value of breakover voltage needed to turn the device on. The schematic symbol of the SUS is shown in Figure 20-4. The breakover voltage value can be adjusted by connecting a zener diode between the gate and cathode, as shown in Figure 20-5. In this example, a 3-volt zener diode is used. This permits the breakover voltage to become about 3.6 volts (3 volts of the zener diode + 0.6 volt needed to turn on any silicon device). The breakover voltage can be adjusted to a value that is less than the original value if no zener diode were to be used, but it cannot be adjusted to a higher value.

UNIT 20 REVIEW QUESTIONS

1. Why was the unijunction transistor developed? _____
_____

2. What is the purpose of the transformer in this circuit?
_____

3. Does the UJT operate on a DC voltage or an AC voltage? _____
_____

4. Is the output from the SCR to the load AC or DC voltage? _____
_____

5. Refer to Figure 20-1. What component is used to vary the pulse rate of the UJT? _____
_____

6. Explain why the UJT can be used to phase shift the SCR. _____
_____
_____

UNIT 21

SCR CONTROL OF A FULL-WAVE RECTIFIER

OBJECTIVES

After studying this unit, the student should be able to:

- Discuss how SCRs can be used to control the DC output voltage of a full-wave rectifier.

- Explain why only two SCRs are needed to control a bridge rectifier.

- Connect a full-wave rectifier circuit controlled by SCRs.

Up to this point, the control of only one SCR in a circuit has been studied. When only one SCR is used, the output voltage is half-wave direct current. Almost all industrial applications for SCRs require full-wave rectification of the AC voltage. SCRs can control the output of a bridge rectifier when connected into a bridge circuit as shown in Figure 21-1.

In this figure, two SCRs have been connected so that their cathodes are tied together to form the positive post of the bridge rectifier. The other two rectifiers are common junction diodes. Because each half of the AC waveform must pass through one of the SCRs, only two are needed for full control of the bridge rectifier.

The phase-shift network can be the same as for a single SCR, Figure 21-2. This circuit can be separated into two sections. One section is the bridge containing the SCRs and the load resistor. The other is the phase-shifting network made up of the step-down transformer, UJT, and related components. This UJT circuit is identical to the circuit used to provide phase-shift control for a single SCR. The only real difference is that this circuit has the gate leads of each SCR tied together through a low value of resistance. These resistors make the gates of the SCRs more closely matched in impedance; one SCR will not fire ahead of the other and take all the gate current. The two low-value resistors force the gates to share the current delivered by the UJT.

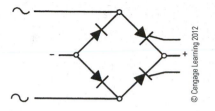

FIGURE 21-1 *Bridge rectifier controlled by SCRs*

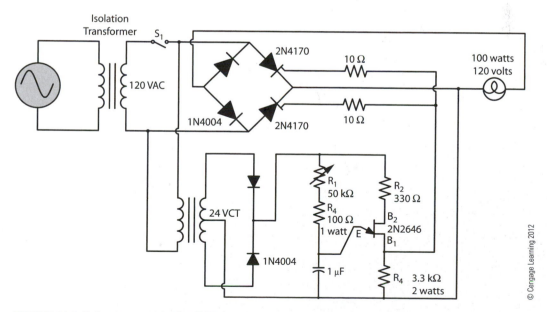

FIGURE 21-2 *Full-wave control using SCRs*

UNIT 21 REVIEW QUESTIONS

1. What type of rectifier are the two SCRs connected into? _____
_____

2. What is the purpose of the two low-value resistors connected to the gates of the SCR? _____
_____
_____

3. What type of rectifier is used to supply DC voltage to the UJT section of the circuit? _____
_____

4. Why are only two SCRs needed to control the output of a full-wave rectifier instead of four? _____
_____
_____

5. What type of output voltage is obtained when only one SCR is used in an AC circuit? _____
_____

UNIT 22

A SOLID-STATE ALARM
WITH BATTERY BACKUP

OBJECTIVES

After studying this unit, the student should be able to:

- Discuss the operation of this alarm system.
- Construct this circuit using discrete components.

The alarm circuit in Figure 22-1 operates on 120 volts AC (120 VAC) during normal operation. If the AC power is interrupted, a 12-volt battery is used to provide continued operation of the circuit. Battery operation continues until the AC power is restored. At this time, the circuit returns to operation on the AC power line. Sensor switches used on doors and windows are normally closed and form one continuous series circuit.

PARTS OF THE ALARM

The parts of this alarm include the following:

- Fuse F_1 is a 1-A buss-type fuse. It is used to protect the circuit against a short circuit.

- Switch S_1 is a double-pole switch. One side of it is used to control the AC voltage to the primary of the transformer. The other side disconnects the 12-volt battery from the DC side of the circuit. The dotted line between the two switches shows mechanical connection. It is the power switch and connects power to the circuit.

- T_1 is a 24-volt center-tapped transformer used to step the 120 VAC down to 24 VAC.

- Diodes D_1 and D_2 are used as rectifiers to change the 24 VAC into DC. Because this is a center-tapped transformer, the DC voltage is about 12 volts.

- Capacitor C_1 is used to filter the output voltage of the rectifier. When the ripple has been filtered, the average DC voltage increases to about 17 volts.

- Diode D_3 is connected in series with the 12-volt battery that supplies the backup power source. Diode D_3 prevents current from being drawn from the battery unless the AC power is interrupted. When the rectifier is operating, 17 volts DC (17 VDC) is applied to the cathode of diode D_3 and 12 volts is applied to its anode. In this condition, diode D_3 is reverse biased and no current flows. If the AC power is interrupted, however, the output of the rectifier turns off. When the voltage of the rectifier drops below 12 volts, diode D_3 becomes forward biased and the battery supplies the power for the continued operation of the circuit.

- Diodes D_4 and D_5 are LEDs used as pilot lights. Diode D_4 indicates that the power switch has been turned on. Diode D_5 indicates that the alarm has been armed.

- Resistors R_1 and R_2 are used to limit the current to D_4 and D_5.

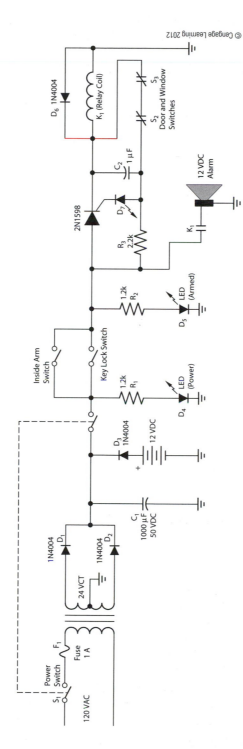

FIGURE 22-1 *Burglar alarm with battery backup*

© Cengage Learning 2012

- The key switch is a key-locked switch installed at an outer door of the building the alarm is to protect. Upon leaving the building, the key switch is closed and the alarm becomes armed. Upon returning to the building, the key switch is opened, disarming the alarm circuit.

- The arm switch is connected in parallel with the key switch and located on the front panel of the alarm. It is used to arm the system from inside the building. (Note: If the alarm is installed inside a home, it can be armed from inside the house at night.)

- The SCR is the real controller of the circuit. It is used because of its characteristic of remaining turned on once the gate has been triggered. When the SCR turns on, it provides the current to turn on relay coil K_1.

- When relay K_1 turns on, its contacts close and connect the alarm bell or siren to the power line. With the SCR turned on, the arm switch or the power switch must be reopened to turn the alarm off.

- Resistor R_3 is used to limit the gate current of the SCR.

- Capacitor C_2 helps prevent false triggering of the SCR due to voltage spikes that may appear on the power line.

- Diode D_6 is used as a kickback or free-wheeling diode to kill the voltage spike induced in the coil of relay K_1 when the power is interrupted.

- Switches S_2 and S_3 are normally closed switches connected to doors, windows, or whatever is to be protected. The schematic shows two of these switches, but any number can be used as long as they are normally closed and connected in series. As long as these switches remain closed, the gate of the SCR is connected to ground and cannot be triggered. If one of them opens, however, it permits current to flow to the gate and trigger the SCR.

- Diode D_7 is an LED connected in series with the gate of the SCR. Its function is to prevent nuisance trips of the alarm. Over time, it is possible for the contacts of switches S_1, S_2, and S_3 to become oxidized and develop resistance across them, increasing the voltage drop to ground. This increased resistance can permit enough current to flow to the gate of the SCR and turn it on. Light-emitting diode D_7 requires a forward voltage drop of approximately 1.7 volts before it will conduct current to the gate of the SCR. This prevents the voltage drop across the normally closed switches due to oxidation from turning on the alarm.

OPERATING THE ALARM

The operation of the circuit is as follows:

1. When switch S_1 is closed, DC voltage is provided by the rectifier. This permits diode D_4 to light to indicate that the power switch is turned on.

2. With either the key switch or the arm switch closed, power is supplied to the SCR. Diode D_5 lights, indicating that the circuit is armed.

3. Because the door and window switches are closed, the gate current of the SCR is connected to the cathode and the SCR will not trigger. If one of them is opened, however, current is provided to trigger the gate of the SCR.

4. When the SCR turns on, relay K_1 turns on and closes the contacts that provide power to the bell or siren. Once the SCR has turned on, it cannot be turned off unless the arm switch or power switch is opened.

If the AC power should fail after the power and arm switch have been turned on, the battery will provide the power for the operation of the circuit. When the AC power is restored, the battery is disconnected from the circuit. This circuit does not require the use of a large battery because it is only used when the AC power is interrupted. A 12-volt lantern battery, which can be purchased at most sporting goods stores, is ideal for this circuit.

UNIT 22 REVIEW QUESTIONS

1. What voltage does this alarm operate on during normal operation? _____

_____

2. What is used to provide backup power in the event of a power failure? _____

_____

3. What type of switch is used for the power switch? _____

_____

4. What is the purpose of diode D_4? _____

_____

5. What is the purpose of diode D_5? _____

_____

6. What is used to arm the system from outside the building? _____

_____

7. Must the door and window switches be open or closed to keep the alarm turned off? _____

_____

8. What is the function of diode D_6? _____

_____

UNIT 23

THE DIAC AND SILICON BILATERAL SWITCH

OBJECTIVES

After studying this unit, the student should be able to:

- Discuss the operation of a diac.

- Draw the schematic symbols of a diac.

- Discuss the operation of a silicon bilateral switch.

- Draw the schematic symbol of an SBS.

- Connect a diac in an electronic circuit.

FIGURE 23-1 *Schematic symbols of a diac*

The **diac** is a special-purpose bidirectional diode. Its primary function is to phase shift a triac. The trigger operation of the diac is similar to that of a unijunction transistor, except the diac is a bi- or two-directional device. It can operate in an AC circuit, whereas the UJT is a DC device only.

There are two schematic symbols for the diac, Figure 23-1. Either symbol is used in an electronic schematic to illustrate the use of a diac, so become familiar with both.

The diac is a voltage-sensitive switch that can operate on either polarity, Figure 23-2. When voltage is applied to the diac, it remains in the turned-off state until the applied voltage reaches a predetermined level. Assume this to be 15 volts. Upon reaching 15 volts, the diac turns on (fires). When the diac fires, it displays a **negative resistance**, meaning that it conducts at a lower voltage than that which was applied to it (assume 5 volts). The diac remains turned on until the applied voltage drops below its conduction level, or 5 volts, Figure 23-3.

Because the diac is a bidirectional device, it conducts on either half cycle of the AC voltage applied to it, Figure 23-4. It has the same operating characteristic with either half cycle of alternating current.

FIGURE 23-2 *Diac operates on either polarity.*

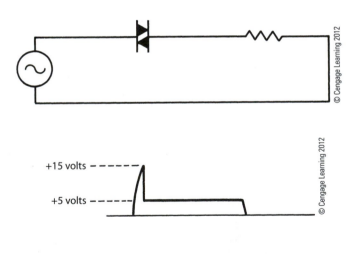

FIGURE 23-3 *The diac conducts at a lower voltage than is required to turn it on.*

FIGURE 23-4 *The diac conducts both halves of the AC waveform.*

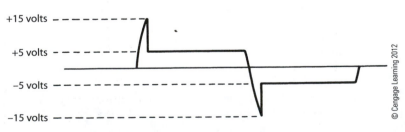

The simplest way to sum up the operation of the diac is to say it is a voltage-sensitive AC switch.

THE SILICON BILATERAL SWITCH

The silicon bilateral switch or SBS is another bidirectional device often used to trigger the gate of a triac. The schematic symbol for an SBS is shown in Figure 23-5. The SBS is very similar to the diac in many ways. Both will conduct current in either direction, and both exhibit negative resistance. The SBS, however, has a more pronounced negative resistance region than does the diac. A characteristic voltage curve for both the diac and the SBS are shown in Figure 23-6. Notice that the break-back voltage of the SBS is much more pronounced than that of the diac. Also, the breakover voltage of an SBS is generally lower than that of a diac. The breakover voltage of a diac will most often range between ±16 volts and ±32 volts. The most common breakover voltage for a silicon bilateral switch is ±8 volts.

The silicon bilateral switch has several other advantages over the diac. SBSs are generally symmetrical to within about 0.3 volt. This means that the positive breakover voltage and the negative breakover voltage will be within about 0.3 volt of each other. The diac is generally symmetrical to within about 1 volt.

The SBS also has a gate lead that permits some control over the breakover voltage. If a 3-volt zener diode were to be connected between the gate and anode 2, as shown in Figure 23-7, the positive breakover voltage would be reduced to about 3.6 volts (3 volts for the zener diode + 0.6 volt needed to turn on a silicon device), but the reverse breakover voltage would be unaffected, as shown in

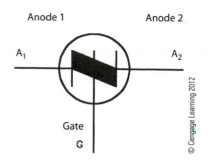

FIGURE 23-5 *Silicon bilateral switch*

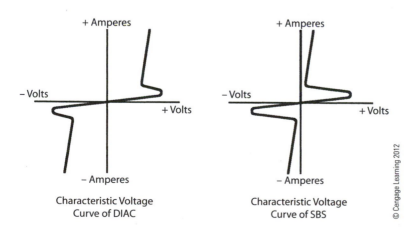

Characteristic Voltage
Curve of DIAC

Characteristic Voltage
Curve of SBS

FIGURE 23-6 *Characteristic voltage curve of diac and SBS*

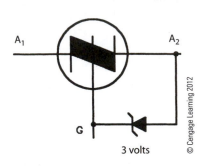

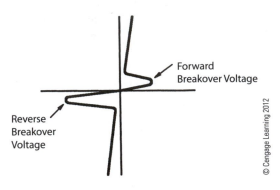

FIGURE 23-7 *Zener diode used to control breakover voltage*

FIGURE 23-8 *Forward and reverse breakover voltages are different.*

Figure 23-8. This could be used if it were desirable to have different forward and reverse breakover values, which is not usually the case. The SBS is most often used with the gate lead not connected.

UNIT 23 REVIEW QUESTIONS

1. What is the primary function of a diac? _____
_____

2. Does the diac operate on AC or DC voltage? _____
_____

3. What characteristic permits the diac to conduct current at a lower voltage than it takes to turn it on?
_____
_____

4. Why can the diac conduct on either half of an AC waveform? _____
_____

5. Make a simple statement that will sum up the operating characteristics of a diac. _____
_____

UNIT 24

THE TRIAC

OBJECTIVES

After studying this unit, the student should be able to:

- Discuss the operation of a triac.
- Draw the schematic symbol of a triac.
- Test a triac with an ohmmeter.
- Connect a triac into a circuit.

The triac is a device very similar in operation to the SCR. Even their appearance is similar, as shown in Figure 24-1, which shows triacs in various case styles. When an SCR is connected into an AC circuit, the output voltage is rectified to direct current. The triac, however, is designed to conduct on both halves of the AC waveform. Therefore, the output of the triac is alternating current instead of direct current.

The triac is made like two SCRs connected in parallel, facing in opposite directions, with their gate leads tied together, Figure 24-2.

OPERATION OF THE TRIAC

Notice that the schematic symbol for the triac, Figure 24-3, is similar to the connection of the two SCRs. The gate must be connected to the same polarity as MT_2 to turn the triac, on. When the voltage applied to MT_2 increases in the positive direction, the gate fires the half of the triac, which is forward biased during that half of the cycle. Because the other SCR half of the triac is reverse biased during that half cycle, it cannot be fired. When the applied AC voltage becomes negative at MT_2, the gate fires the other half of the triac, which is now forward biased. This action can be seen with an oscilloscope, Figure 24-4.

Adding a Diode

If a diode is connected in series with the gate, it permits only half of the AC cycle to provide gate current. The gate provides current flow for only half of the AC cycle: assume the positive half cycle. If

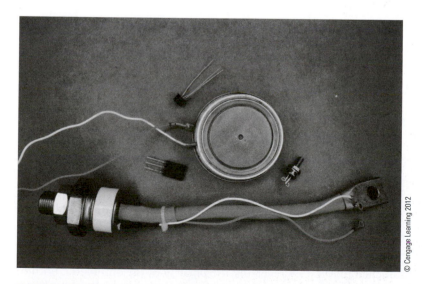

FIGURE 24-1 *Triacs shown in various sizes and case sizes*

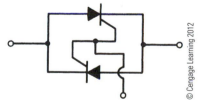

FIGURE 24-2 *The triac operates like two SCRs connected together.*

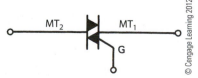

FIGURE 24-3 *Schematic symbol of a triac*

FIGURE 24-4 *The diode permits only one-half of the triac to conduct.*

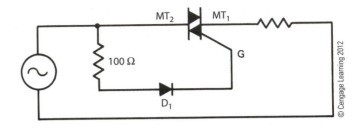

FIGURE 24-5 *Gate current controlled by a variable resistor*

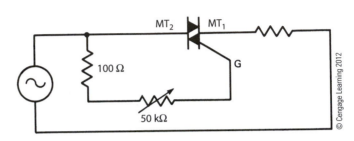

an oscilloscope is connected across the load resistor, only the positive half of the AC waveform is seen. The diode is reverse biased during the negative half cycle, and no gate current flows to turn the negative half of the triac on.

If the diode is removed from the circuit and replaced facing in the opposite direction, it blocks the positive half cycle and passes the negative half. The oscilloscope connected across the load resistor shows the negative half of the AC waveform. This is a good example of the two-SCR arrangement of the triac.

Adding a Variable Resistor

If a variable resistor is used to control the gate of the triac, the same control problems exist as did with the SCR. A simple resistor control, as shown in Figure 24-5, permits the triac to control only half of the waveform just like an SCR that has not been phase shifted.

TESTING THE TRIAC

The triac can be tested with an ohmmeter; because it is basically two SCRs, Figure 24-2, it can be tested the same way. To test the triac, connect the meter leads of the ohmmeter to MT_1 and MT_2 of the triac. The meter should indicate no continuity until the gate lead is touched to MT_2. One of the SCRs in the triac has now been tested. To test the other, reverse the meter leads connected to MT_1 and MT_2. No continuity should be read until the gate lead is again touched to MT_2. By reversing the polarity of the ohmmeter, both of the SCRs in the triac can be tested.

APPLICATIONS

The ability of a triac to control a large amount of load current with a small amount of gate current can be seen in the circuit shown in Figure 24-6. In this circuit, a small 24-volt thermostat is used to control the operation of a fan motor. It is assumed that the maximum current rating of the thermostat contacts is 0.5 A, and the full load running current of the motor is 7.5 A. If the thermostat contacts were to be used to control the operation of the motor, they would be destroyed the first time the motor was turned on.

If a triac is used as the controlling element for the motor, however, the thermostat can easily furnish the amount of gate current necessary to control the triac. A 24-volt step-down transformer is used to reduce the voltage in the thermostat circuit. The resistor limits the amount of current flow in the gate–MT_1 circuit when the thermostat contacts close. Notice that one side of the transformer secondary is connected directly to MT_1. This must be done to provide a complete path for current flow when the thermostat contacts close.

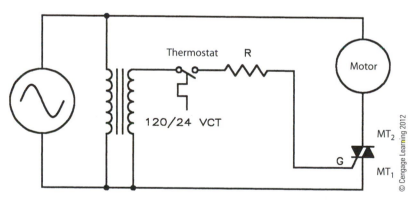

FIGURE 24-6 *Triac controls motor operation.*

UNIT 24 REVIEW QUESTIONS

1. When a triac is connected into an AC circuit, is the output voltage AC or DC? _____

_____

2. The gate of the triac must be connected to the same polarity as which other lead for it to turn on?

_____

_____

3. If a diode is connected in the gate circuit of a triac, will the output voltage of the triac be AC or DC?

_____

_____

4. When testing a triac with an ohmmeter, will the triac conduct with the meter leads connected to MT$_1$ and MT$_2$ in either direction or only one direction? _____

_____

5. How would two SCRs be connected to give similar operating characteristics as a triac?

_____

_____

UNIT 25

PHASE SHIFTING
THE TRIAC

OBJECTIVES

After studying this unit, the student should be able to:

- Discuss phase shift control for a triac.

- Construct a triac circuit using discrete components.

- Make electrical measurements using test equipment.

The triac, like the SCR, must be phase shifted if complete control is to be gained over the AC voltage applied to it. To phase shift the triac, separate the gate pulses from the voltage applied to MT_1 and MT_2. This is the same method used with the SCR.

One method of phase shifting the SCR is to use the unijunction transistor to supply pulses to the gate. Because the UJT is a DC-operated device, it cannot be used to supply the trigger pulses for the triac, which needs both negative and positive pulses. The device most often used to phase shift the triac is the diac, which is a bipolar device. It can supply both the negative and positive pulses needed to trigger the triac, Figure 25-1.

Because the diac operates like an AC voltage-sensitive switch, assume that it turns on when the voltage applied to it reaches 30 volts and turns off when the voltage drops to 5 volts. In the operation of the circuit in Figure 25-1, capacitor C_1 is charged through resistors R_1 and R_2. The diac remains off until the voltage of capacitor C_1 reaches 30 volts. When C_1 reaches this value, the diac turns on and discharges C_1 through the gate of the triac. This discharge pulse fires the triac. When the charge on the capacitor drops to 5 volts, the diac turns off again.

The phase-shift circuit in Figure 25-1 is connected in parallel with the triac. Because the triac is essentially two SCRs, when it fires in one direction the voltage drop across the device becomes about 1 volt. It remains there until the AC wave drops back to 0 and turns the device off. Once the diac fires and discharges capacitor C_1, the capacitor cannot begin to charge again until the triac has turned off at the end of the AC cycle. As long as the triac is conducting, there is only about 1 volt applied to the phase-shift circuit.

When capacitor C_1 again begins to charge, it will charge at the opposite polarity from the preceding cycle. The diac turns on when the voltage across C_1 reaches 30 volts and discharges the capacitor through the gate of the triac. When the triac fires, it conducts the

FIGURE 25-1 *Diac used to phase shift a triac*

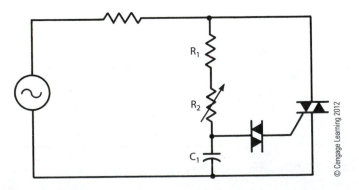

© Cengage Learning 2012

other half of the AC waveform. The voltage applied to the phase-shift circuit drops to about 1 volt as long as the triac is conducting. The capacitor cannot charge until the triac turns off at the end of that half cycle of alternating current. The triac can be triggered by either half cycle of the AC voltage applied to it.

The charge time of the capacitor is determined by the size of C_1 and the amount of resistance connected in series with it. By varying the resistance of R_2, the time it takes C_1 to charge to 30 volts can be adjusted to different lengths. This allows the triac to be fired at different points along the cycle of AC voltage connected to it, which is the requirement for phase shifting a thyristor.

APPLICATIONS

A good example of how a phase-shifted triac can be used is seen in Figure 25-2. In this circuit, a triac controls the amount of charging current to a battery. The triac controls the current flow through the

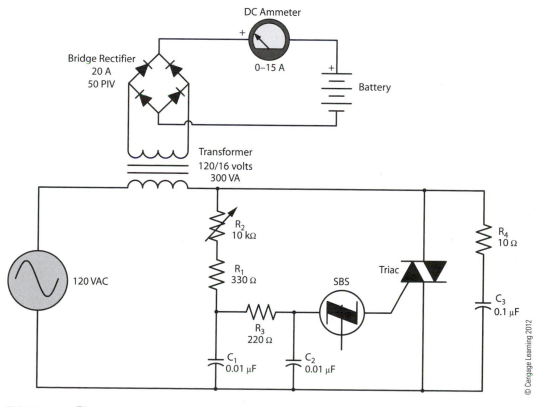

FIGURE 25-2 *Triac-controlled battery charging circuit*

© Cengage Learning 2012

primary winding of a transformer, which in turn controls the transformer's secondary current. A bridge rectifier converts the alternating current delivered by the secondary of the transformer into direct current.

Components R_1, R_2, R_3, C_1, and C_2 are part of the phase-shift network for the triac. Resistor R_2 is variable, which permits the amount of current flow through the primary of the transformer to be controlled. The value of resistor R_1 sets an upper limit on the amount of charging current that can be delivered to the battery. A silicon bilateral switch has been connected in series with the gate of the triac. The negative resistance characteristic of the SBS is used to permit capacitors C_1 and C_2 to provide a pulse of current to the gate of the triac. Resistor R_3 slightly extends the discharge rate of capacitor C_1, which provides a current pulse of longer duration to the gate of the triac. Resistor R_4 and capacitor C_3 provide transient or spike voltage protection for the circuit.

UNIT 25 REVIEW QUESTIONS

1. What must be done to phase shift the triac? _____
   _____

2. What electronic component is generally used to phase shift the triac? _____
   _____

3. Why is it necessary to phase shift the triac? _____
   _____

4. Refer to the circuit shown in Figure 23-1 in Laboratory Exercise 23. What determines the amount
   of time between pulses to the gate of the traic? _____
   _____

5. What is the function of resistor R_1 in the circuit shown in Figure 23-1 in Laboratory Exercise 23?
   _____
   _____

UNIT 26

OTHER METHODS OF
AC VOLTAGE CONTROL

OBJECTIVES

After studying this unit, the student should be able to:

- Discuss different methods of controlling an AC voltage.

- Construct a circuit for controlling AC voltage using a bridge rectifier and an SCR circuit.

- Construct a circuit for controlling AC voltage using a transistor and a bridge rectifier.

An AC voltage can be controlled by electronic devices, the most common of which is the triac. There are other devices that can be used to control an AC voltage when properly connected, for example the SCR. It can be used as an AC switch when connected in a circuit as shown in Figure 26-1.

THE SCR FOR VOLTAGE CONTROL

In this circuit, a bridge rectifier is connected in series with the load resistor. If current is to flow to the load, it must pass through the SCR, which is connected across the positive and negative points of the bridge rectifier.

Operating the Circuit with the SCR

In order to understand the operation of this circuit:

1. Assume that point X of the AC line is positive and point Y is negative, also that current flow is from positive to negative.

2. Current flows to point A of the rectifier.

3. Diode D_1 is forward biased, so current flows to point B of the rectifier.

4. At point B, diode D_2 is reverse biased, so current flows through the SCR to point D of the rectifier.

5. At point D, both diodes D_3 and D_4 are forward biased, and current does not flow from positive to positive.

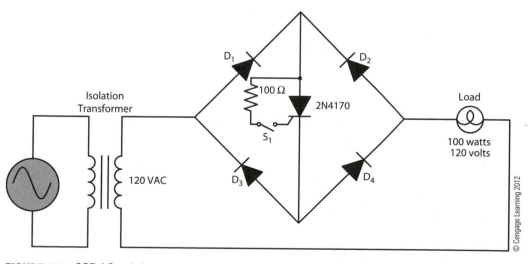

FIGURE 26-1 *SCR AC switch*

6. The current flows through diode D_4 to point C of the rectifier.

7. From here, the current flows through the load resistor to the other side of the AC line at point Y.

8. Because current flowed through the load resistor during the half cycle, a positive voltage appears across the load resistor.

Now, assume that point Y of the AC line has become positive and point X is now negative.

1. Current flows from point Y, through the load resistor to point C of the bridge.

2. The current flows through diode D_2 to point B.

3. From point B, current flows through the SCR to point D of the rectifier.

4. Both diodes D_3 and D_4 are forward biased, so the current flows to the most negative point, or through diode D_3 to point X of the AC line.

5. Because current flowed through the load resistor during the half cycle, a voltage appears across the resistor.

6. The current, however, flowed through the load resistor in the opposite direction. Therefore, a negative voltage appears across the resistor.

Notice that the current flowed through the load resistor in both directions, but it flowed through the SCR in only one direction. This type of circuit permits a DC device to control alternating current.

THE TRIAC FOR VOLTAGE CONTROL

The triac is most often used to control alternating current, but there are some conditions under which a triac can have undesirable characteristics. The triac operates like two SCRs connected in opposite directions, Figure 26-2. If the internal structure of the triac is unbalanced, it causes one section to conduct before the other. The positive half of the AC wave will start conducting before the negative half. Figure 26-3 shows an oscilloscope that has been connected across the load resistor. The variable resistor is adjusted to permit current to flow through the load resistor. The half of the triac that controls the positive half of the AC wave begins to conduct before the half that controls the negative portion. A waveform similar to the one shown in Figure 26-4 is seen on the display of the oscilloscope. The waveform in Figure 26-4 shows that only a portion of the positive half cycle is being conducted. The load resistor, therefore, has DC voltage applied to it.

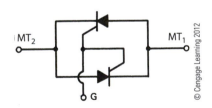

FIGURE 26-2 *A triac operates like two SCRs connected together.*

© Cengage Learning 2012

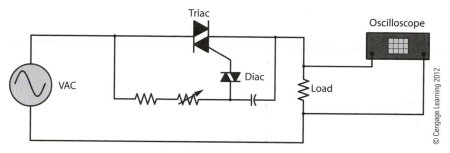

FIGURE 26-3 *Triac control of AC voltage*

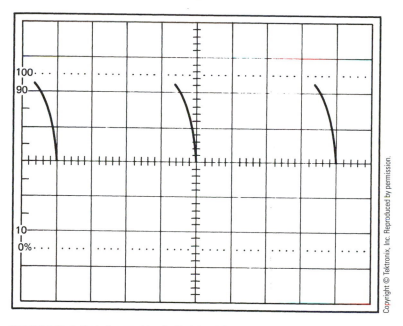

FIGURE 26-4 *Only the positive half of the AC waveform is being conducted.*

LIMITATIONS TO CONTROL OF THE CIRCUIT

If DC voltage is applied to a pure resistance, no adverse conditions will occur. If DC voltage is applied to a load that is inductive, however, a great deal of damage can be done. When an AC voltage is applied to an inductive load, the current is limited by inductive reactance; with DC voltage, however, the current is limited only by the wire resistance of the coil. Most AC inductive loads such as motors or transformers have a low wire resistance. By applying DC voltage to one of these loads, the motor or transformer can be destroyed.

Solving the Problem

When an electronic device is used to control an AC voltage, the device must conduct both the negative and positive halves of the AC

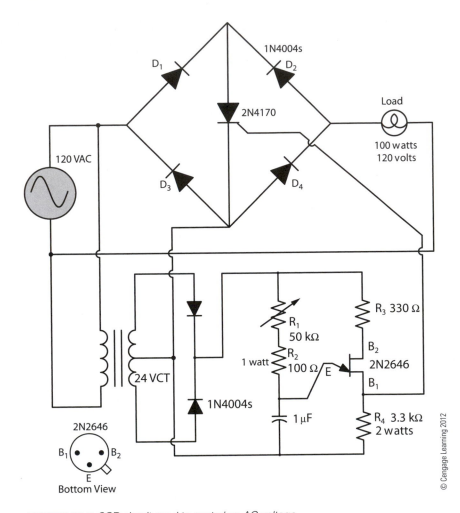

FIGURE 26-5 *SCR circuit used to control an AC voltage*

waveform, especially if the AC voltage is applied to an inductive load. If a phase shift circuit is added to the SCR shown in Figure 26-1, both the negative and positive halves of the AC waveform will be conducted. In the circuit shown in Figure 26-5, the SCR is controlling a DC voltage. It begins conducting at the same point for each DC pulse applied to it. Because one of the DC pulses permits current to flow through the load resistor in one direction, and the next pulse permits current to flow through the load resistor in the opposite direction, the SCR permits both the positive and negative halves of the AC waveform to be conducted.

If an oscilloscope is connected to the load resistor in the circuit shown in Figure 26-5, a waveform similar to the one shown in

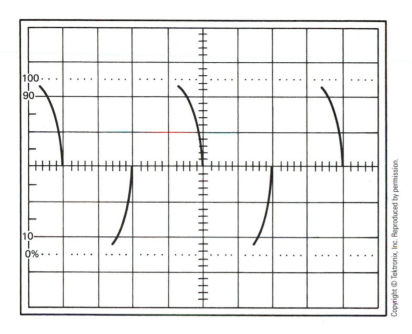

FIGURE 26-6 *Both halves of the AC waveform are being conducted.*

Figure 26-6 can be seen. Notice that both the positive and negative halves of the AC waveform have been conducted.

The AC output voltage of the circuit in Figure 26-5 is controlled by permitting the SCR to fire at different points in a cycle. The voltage applied to the load is determined by the amount of time the SCR is conducting compared to the amount of time it is not.

OTHER DEVICES USED FOR CONTROL

Thyristor devices such as the SCR and triac control the voltage by chopping the waveform. Resistance loads or motors operate without problems when this type of waveform is applied to them. Other devices, however, will not. If a variable AC voltage is applied to these devices, it must be done by increasing or decreasing the amplitude of the waveform. This is accomplished by replacing the SCR with a transistor, Figure 26-7. The transistor operates by permitting more or less current to flow through it, not by turning on or off at different points in the waveform. Because the transistor controls the AC voltage applied to the load by controlling the current flow, voltage is dropped across the collector–emitter of the transistor. Care must be taken not to exceed the transistor's voltage, current, or power ratings.

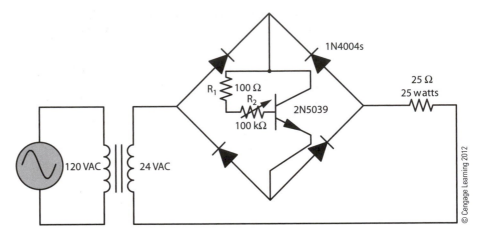

FIGURE 26-7 *AC voltage controlled by a transistor*

APPLICATIONS

As previously stated, when variable AC voltage is to be connected to an induction motor to perform the job of speed control, it is very important that both halves of the AC waveform are conducted to the stator of the motor. The circuit shown in Figure 26-8 can be used to provide variable voltage to the windings of an AC induction motor. A bridge rectifier has been connected in series with the motor windings. An SCR is used to control the current flow through the bridge rectifier. Gate current for the SCR is provided by a circuit consisting

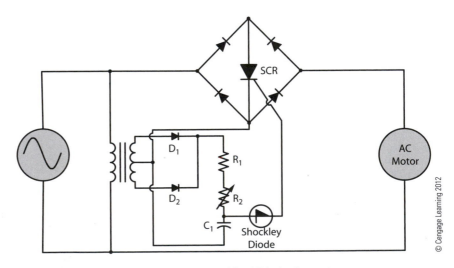

FIGURE 26-8 *Variable voltage control for AC induction motor*

of a step-down transformer with a center-tapped secondary winding, diodes D_1 and D_2, resistors R_1 and R_2, capacitor C_1, and a Shockley diode.

The transformer is used to reduce the AC line voltage from 120 to 24 volts. Diodes D_1 and D_2 form a two-diode type of bridge rectifier to provide full-wave rectification for the gate of the SCR. Resistor R_1 limits current to the gate of the SCR if the value of resistor R_2 should be adjusted to 0. The adjustment of resistor R_2 determines at what point in the circuit the SCR will turn on. Capacitor C_1 serves two functions in this circuit. It produces a time delay by forming an RC time constant with resistors R_1 and R_2. It also provides a pulse of current to the gate of the SCR when the Shockley diode turns on and permits current to flow. The negative resistance characteristic of the Shockley diode is used to permit the SCR to be phase shifted.

UNIT 26 REVIEW QUESTIONS

1. When an SCR and bridge rectifier are used to control an AC voltage, how is the bridge rectifier connected in respect to the load? _____

_____

2. Explain how it is possible for a DC device such as an SCR or transistor to control an AC voltage.

_____

_____

3. Explain why a triac is sometimes an undesirable device for controlling the AC voltage applied to an inductive device. _____

_____

4. Explain why care should be taken to not permit a DC voltage to be applied to an inductive load.

_____

_____

5. What is the advantage of using the transistor as the control device instead of the SCR?

_____

_____

THE SOLID-STATE RELAY

OBJECTIVES

After studying this unit, the student should be able to:

- Discuss the operation of a solid-state relay.

- Discuss different methods used to isolate the control section of the relay from the power section.

- Connect a solid-state relay in an electrical circuit.

- Discuss different uses and applications for solid-state relays.

The solid-state relay is a device that has become increasingly popular for switching applications. The solid-state relay has no moving parts, is resistant to shock and vibration, and is sealed against dirt and moisture. Its greatest advantage, however, is the fact that the control input voltage is isolated from the line device the relay is intended to control, Figure 27-1.

Solid-state relays can be used to control either a DC or an AC load. If the relay is designed to control a DC load, a power transistor connects the load to the line, as shown in Figure 27-2.

THE LED IN THE RELAY

This relay has an LED connected to the input or control voltage. When the input voltage turns the LED on, a photodetector connected to the base of the transistor turns the transistor on and connects the load to the line. This optical coupling is commonly used with solid-state relays. These relays are referred to as being optoisolated. This means that the load side of the relay is optically isolated from the control side of the relay. The control medium is a light beam. No voltage spikes or electrical noise produced on the load side of the relay are therefore transmitted to the control side.

THE TRIAC IN THE RELAY

Solid-state relays intended for use as AC controllers have a triac connected to the load circuit in place of a power transistor. In Figure 27-3, an LED is used as the control device just as in the

FIGURE 27-1 *Solid-state relay*

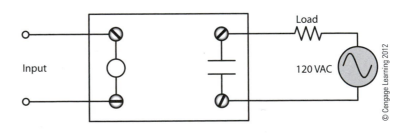

FIGURE 27-2 *Solid-state relay used to control a DC load*

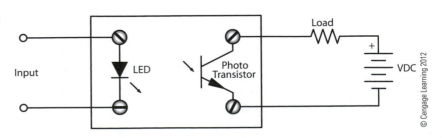

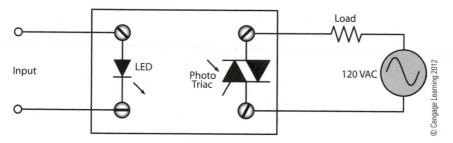

FIGURE 27-3 *Solid-state relay used to control an AC load*

previous example. When the photodetector sees the LED, it triggers the gate of the triac and connects the load to the line.

Other Devices in the Relay

Although optoisolation, Figure 27-4, is probably the most common method used for the control of a solid-state relay, it is not the only method used. Some relays use a small reed relay to control the output, Figure 27-5. A small set of reed contacts are connected to the gate of the triac. The control circuit is connected to the coil of the reed relay. When the control voltage causes a current to flow through the coil, a magnetic field is produced around the coil of the relay. This magnetic field closes the reed contacts and the triac turns on. In this type of solid-state relay, a magnetic field, instead of a light beam, isolates the control circuit from the load circuit.

Control Voltage

The control voltage for most solid-state relays ranges from about 3 to 32 volts and can be direct current or alternating current. If a triac is used as the control device, load voltage ratings of 120–240 volts AC are common. Current ratings can range from 5 to 25 A. Many solid-state relays have a feature called zero switching. Assume that the AC voltage is at its positive peak value when the gate tells the triac to turn off. The triac continues to conduct until the AC voltage drops to a 0 level before it actually turns off.

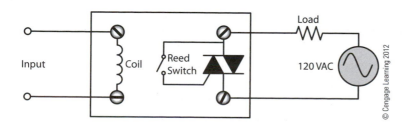

FIGURE 27-4 *Reed relay used as a triac driver*

FIGURE 27-5 *Control circuit isolated by a reed relay*

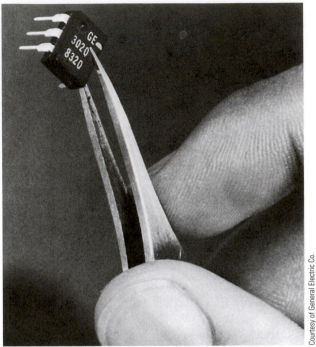

Courtesy of General Electric Co.

VARIOUS TYPES OF RELAYS

Solid-state relays are available in different case styles and power ratings; a solid-state relay is shown in Figure 27-6. Some solid-state relays are designed to be used as time-delay relays.

One of the most common uses for the solid-state relay is the I/O track of a programmable logic controller, Figure 27-7. This controller is basically a computer programmed to perform the same job as a control circuit using relays. Most of the actual control logic is performed by internal relays. These are digital-logic circuits programmed to operate like relays with normally open or normally closed contacts. Although the control logic is performed inside the computer, the computer must be able to communicate with the outside world to accomplish anything. The computer section of the programmable logic controller receives input information from sensing devices: float, pressure, and limit switches, as well as push buttons, Figure 27-8. The computer is also able to give commands to the outside circuits to start and stop motors or open and close valves.

The computer section of the programmable logic controller operates on a DC voltage that generally ranges from 5 to 15 volts, depending on the type of controller. The voltage source must be well filtered, registered, and free of voltage spikes and electrical noise. If

FIGURE 27-6 *Solid-state relay*

FIGURE 27-7 *I/O track of a programmable logic controller*

a voltage spike reaches the computer, it can be interpreted as a command by the internal logic. For this reason, there must be electrical isolation between the computer and the circuits outside it.

The function of the I/O track is to provide communication between the computer and the outside circuits while maintaining

FIGURE 27-8 *Central processor unit*

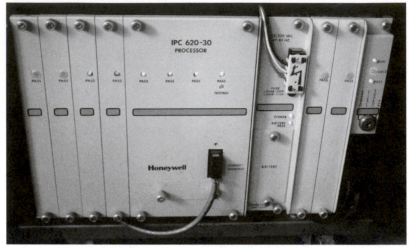

© Cengage Learning 2012

FIGURE 27-9 *Float switch connected to an input module*

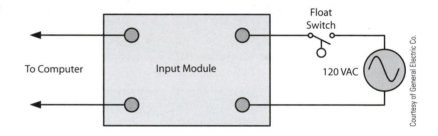

Courtesy of General Electric Co.

isolation between the two. Solid-state relays are used to perform this task. The I/O track contains both input and output modules.

An input module supplies information to the computer from an outside device. The circuit in Figure 27-9 shows a float switch connected to an input module. When the switch closes, 120 volts AC is connected to the input of the solid-state relay. The relay provides a signal to the computer to tell it that the float switch has closed. This example shows 120 volts AC used as the input voltage. In actual practice, however, different types of programmable logic controllers require a different voltage. One unit may use low voltage AC whereas another may use low voltage DC.

Input modules generally contain spike or surge suppression in the form of an electronic device called an metal oxide varistor (MOV). Metal oxide varistors are a type of thyristor that exhibit a change of resistance with a change of voltage. They will have a very high resistance, generally several hundred thousand ohms, until the voltage across them reaches a threshold. When the threshold voltage is reached, the resistance suddenly changes to a very low value of 2 or 3 Ω. This sudden reduction of resistance effectively shorts the circuit and prevents the voltage from going any higher. MOVs are extremely fast acting. They typically change from a very high to a very low value in a few picoseconds. The schematic symbol for an MOV is shown in Figure 27-10.

The MOV is a bidirectional device, meaning that it can conduct current in either direction. This gives it the ability to be used in either DC or AC circuits. They are typically connected across the coils of relays and starters in motor control systems to help prevent voltage spikes from being inducted into the control circuit by collapsing magnetic fields. They are also used in the surge suppressors protecting computers and other types of sensitive electronic equipment. A typical schematic for a surge suppressor is shown in Figure 27-11. Series inductors limit the rise time of current in the circuit and the capacitor limits the rise time of voltage. These elements permit the MOV time to turn on and stop the voltage spike from becoming excessive.

The voltage rating of the MOV will be slightly greater than the normal system voltage. A 120-VAC circuit will generally be protected by an MOV with an AC voltage rating of 140 volts. As long as the voltage remains below 140 volts, the MOV exhibits a very high resistance. When the voltage reaches 140 volts, the MOV suddenly turns on and prevents the voltage from going higher. A typical sine wave and voltage spike is shown in Figure 27-12. A typical input I/O circuit is shown in Figure 27-13. In this circuit, the MOV is used to prevent any voltage spikes from reaching the rectifier or LED.

The output module is used to permit the computer to communicate with the outside circuit. The circuits in Figure 27-14 show the computer controlling the relay coil of a large motor starter.

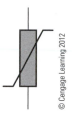

FIGURE 27-10 *Schematic symbol for a metal oxide varistor (MOV)*

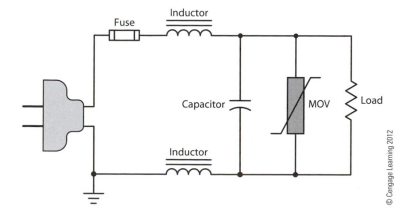

FIGURE 27-11 *Typical surge suppressor circuit using an MOV*

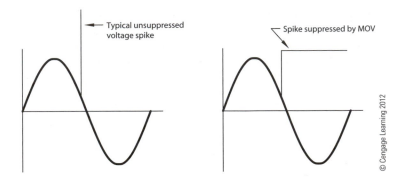

FIGURE 27-12 *Unsuppressed and suppressed voltage spikes*

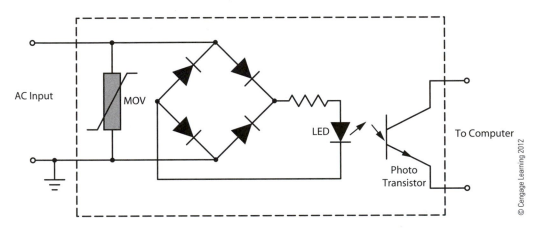

FIGURE 27-13 *MOV protects input I/O of a programmable logic controller.*

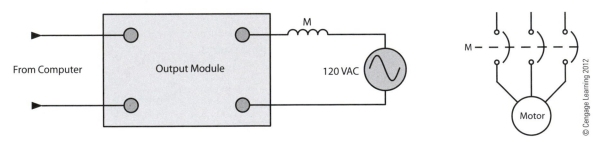

FIGURE 27-14 *Output module controls relay coil.*

The motor starter connects the motor to the line. If the load is small enough, the solid-state relay is used as the controller without an intermediate device. A solenoid valve, for example, could be controlled directly by the output module.

These examples show how solid-state relays can be used. The advantages of the solid-state relay are evident, and an increase in their use is expected.

UNIT 27 REVIEW QUESTIONS

1. When a solid-state relay is used to control an AC load, what electronic device is used to connect the load to the line? _____

_____

2. When a solid-state relay is used to control a DC load, what electronic device is used to connect the load to the line? _____

_____

3. What does optoisolation mean? _____

_____

4. Name two methods of control for solid-state relays. _____

_____

5. Does the output of the solid-state relay provide power to operate the load connected to it?

_____

_____

UNIT 28

THE OSCILLATOR

OBJECTIVES

After studying this unit, the student should be able to:

- Discuss the operation of an oscillator.

- Discuss different uses for the oscillator.

- Construct a square-wave oscillator using discrete electronic components.

Oscillators are devices used to change direct current into alternating current. Their power ratings range from a few milliwatts to megawatts and their frequency from a few cycles to several billion cycles per second. Oscillators are used in hundreds of applications from providing the small signal pulses for operating computer logic circuits to the heating of metal by magnetic induction.

CONSTRUCTING THE OSCILLATOR

There are several ways an oscillator can be constructed. The design is determined by several factors: the amount of power it must produce, its operating frequency, and how stable the frequency must be. Industry is making use of more and more oscillators on the production line, so the maintenance electrician should become familiar with their basic principles of operation.

APPLICATIONS OF THE OSCILLATOR

One of the most common applications of oscillators in industry is the speed control of AC induction motors. Two things determine the speed of an AC induction motor: the number of stator poles and the frequency. By changing the frequency applied to the motor, the speed of the rotating magnetic field (the **synchronous speed**) can be changed also. For instance, if a four-pole induction motor is operated at 60 Hz, the synchronous speed is 1800 revolutions per minute (RPM). If the frequency is reduced to 30 Hz, the synchronous speed will be 900 rpm. As the frequency is reduced in value, however, the inductive reactance of the motor winding is also reduced. The reduction of inductive reactance (impedance) is dealt with, however, by reducing the applied voltage as the frequency is reduced. This reduction prevents the current from becoming excessive in the stator winding and damaging the motor.

Most variable frequency drives operate by first changing the AC voltage into DC and then changing it back to AC at the desired frequency. A variable frequency drive is shown in Figure 28-1. There are several methods used to change the DC voltage back into AC. The method employed is generally determined by the manufacturer, age of the equipment, and the size of motor the drive must control. Variable frequency drives intended to control the speed of motors up to 500 Hp generally use transistors. In the circuit shown in Figure 28-2, a three-phase bridge changes the three-phase alternating current into direct current. The bridge rectifier uses silicon-controlled rectifiers (SCRs) instead of diodes. The SCRs permit the output voltage of the rectifier to be controlled. As the frequency decreases, the SCRs fire later in the cycle and lower the output voltage to the transistors. A choke coil and capacitor bank are used to filter the output voltage before transistors

FIGURE 28-1 *Two-horsepower variable frequency drive*

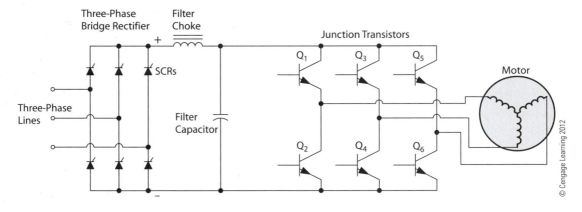

FIGURE 28-2 *Three-phase oscillator*

Q_1 through Q_6 change the DC voltage back into AC. A variable frequency drive using transistors in the output stage is shown in Figure 28-3. An electronic control unit is connected to the bases of transistor Q_1 through Q_6. The control unit converts the DC voltage back into three-phase alternating current by turning transistors on or off at the proper time and in the proper sequence. Assume, for example, that transistors Q_1 and Q_4 are switched on at the same time. This permits stator winding T_1 to be connected to a positive voltage and T_2 to be connected to a negative voltage. Current can flow through Q_4 to T_2 through the motor stator winding and through T_1 to Q_1.

Now assume that transistors Q_1 and Q_4 are switched off and transistors Q_3 and Q_6 are switched on. Current will now flow through

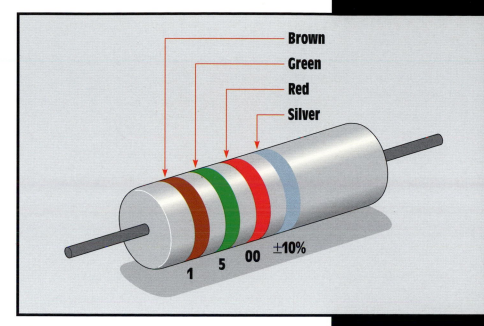

Determining resistor values using the color code

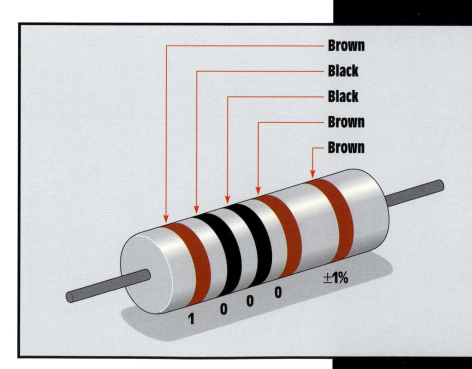

Determining the value of a ±1% resistor

TUBULAR CAPACITORS

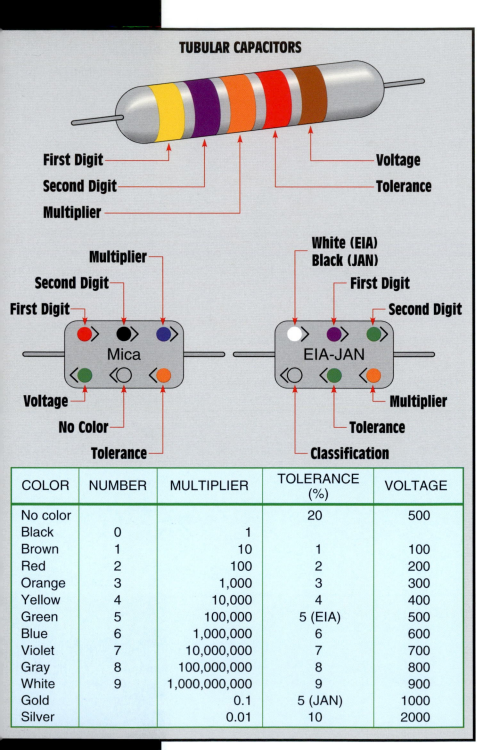

COLOR	NUMBER	MULTIPLIER	TOLERANCE (%)	VOLTAGE
No color			20	500
Black	0	1		
Brown	1	10	1	100
Red	2	100	2	200
Orange	3	1,000	3	300
Yellow	4	10,000	4	400
Green	5	100,000	5 (EIA)	500
Blue	6	1,000,000	6	600
Violet	7	10,000,000	7	700
Gray	8	100,000,000	8	800
White	9	1,000,000,000	9	900
Gold		0.1	5 (JAN)	1000
Silver		0.01	10	2000

Identification of mica and tubular capacitors

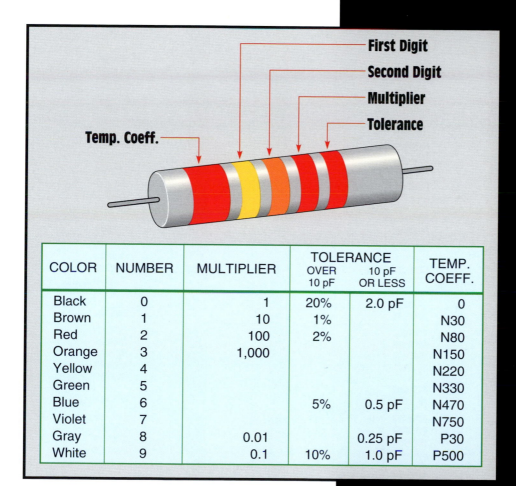

COLOR	NUMBER	MULTIPLIER	TOLERANCE OVER 10 pF	10 pF OR LESS	TEMP. COEFF.
Black	0	1	20%	2.0 pF	0
Brown	1	10	1%		N30
Red	2	100	2%		N80
Orange	3	1,000			N150
Yellow	4				N220
Green	5				N330
Blue	6		5%	0.5 pF	N470
Violet	7				N750
Gray	8	0.01		0.25 pF	P30
White	9	0.1	10%	1.0 pF	P500

Color codes for ceramic capacitors

Left Table

First Band	Second Band	Multiplier Band	Resistance Ω
Brown	Black	Black	10
		Brown	100
		Red	1000
		Orange	10000
		Yellow	0.1 MΩ
		Green	1.0 MΩ
		Blue	1.0 MΩ
	Brown	Black	11
		Brown	110
		Red	1100
		Orange	11000
		Yellow	0.11 MΩ
		Green	1.1 MΩ
		Blue	11.0 MΩ
	Red	Black	12
		Brown	120
		Red	1200
		Orange	12000
		Yellow	0.12 MΩ
		Green	1.2 MΩ
		Blue	12.0 MΩ
	Orange	Black	13
		Brown	130
		Red	1300
		Orange	13000
		Yellow	0.13 MΩ
		Green	1.3 MΩ
		Blue	13.0 MΩ
	Green	Black	15
		Brown	150
		Red	1500
		Orange	15000
		Yellow	0.15 MΩ
		Green	1.5 MΩ
		Blue	15.0 MΩ
	Blue	Black	16
		Brown	160
		Red	1600
		Orange	16000
		Yellow	0.16 MΩ
		Green	1.6 MΩ
		Blue	16.0 MΩ
	Gray	Black	18
		Brown	180
		Red	1800
		Orange	18000
		Yellow	0.18 MΩ
		Green	1.8 MΩ
		Blue	18.0 MΩ

Right Table

First Band	Second Band	Multiplier Band	Resistance Ω
Yellow	Orange	Gold	4.3
		Black	43
		Brown	430
		Red	4300
		Orange	43000
		Yellow	0.43 MΩ
		Green	4.3 MΩ
	Violet	Gold	4.7
		Black	47
		Brown	470
		Red	4700
		Orange	47000
		Yellow	0.47 MΩ
		Green	4.7 MΩ
Green	Brown	Gold	5.1
		Black	51
		Brown	510
		Red	5100
		Orange	51000
		Yellow	0.51 MΩ
		Green	5.1 MΩ
	Blue	Gold	5.6
		Black	56
		Brown	560
		Red	5600
		Orange	56000
		Yellow	0.56 MΩ
		Green	5.6 MΩ
Blue	Red	Gold	6.2
		Black	62
		Brown	620
		Red	6200
		Orange	62000
		Yellow	0.62 MΩ
		Green	6.2 MΩ
	Gray	Gold	6.8
		Black	68
		Brown	680
		Red	6800
		Orange	68000
		Yellow	0.68 MΩ
		Green	6.8 MΩ
Violet	Green	Gold	7.5
		Black	75
		Brown	750
		Red	7500
		Orange	75000
		Yellow	0.75 MΩ
		Green	7.5 MΩ
Gray	Red	Gold	8.2
		Black	82
		Brown	820
		Red	8200
		Orange	82000
		Yellow	0.82 MΩ
		Green	8.2 MΩ
White	Brown	Gold	9.1
		Black	91
		Brown	910
		Red	9100
		Orange	91000
		Yellow	0.91 MΩ
		Green	9.1 MΩ

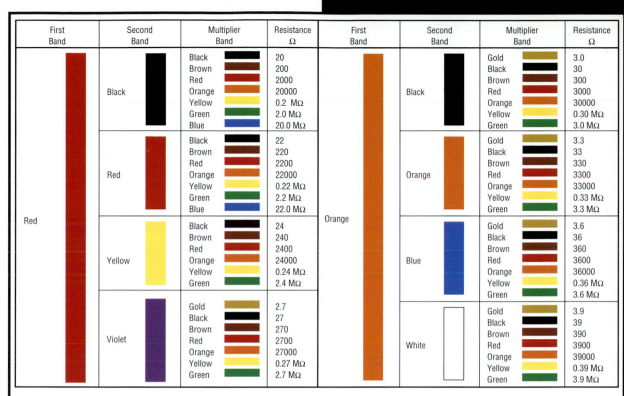

First Band	Second Band	Multiplier Band	Resistance Ω	First Band	Second Band	Multiplier Band	Resistance Ω
Red	Black	Black	20	Orange	Black	Gold	3.0
		Brown	200			Black	30
		Red	2000			Brown	300
		Orange	20000			Red	3000
		Yellow	0.2 MΩ			Orange	30000
		Green	2.0 MΩ			Yellow	0.30 MΩ
		Blue	20.0 MΩ			Green	3.0 MΩ
	Red	Black	22		Orange	Gold	3.3
		Brown	220			Black	33
		Red	2200			Brown	330
		Orange	22000			Red	3300
		Yellow	0.22 MΩ			Orange	33000
		Green	2.2 MΩ			Yellow	0.33 MΩ
		Blue	22.0 MΩ			Green	3.3 MΩ
	Yellow	Black	24		Blue	Gold	3.6
		Brown	240			Black	36
		Red	2400			Brown	360
		Orange	24000			Red	3600
		Yellow	0.24 MΩ			Orange	36000
		Green	2.4 MΩ			Yellow	0.36 MΩ
						Green	3.6 MΩ
	Violet	Gold	2.7		White	Gold	3.9
		Black	27			Black	39
		Brown	270			Brown	390
		Red	2700			Red	3900
		Orange	27000			Orange	39000
		Yellow	0.27 MΩ			Yellow	0.39 MΩ
		Green	2.7 MΩ			Green	3.9 MΩ

Standard Color Code

*Note: Wide Space

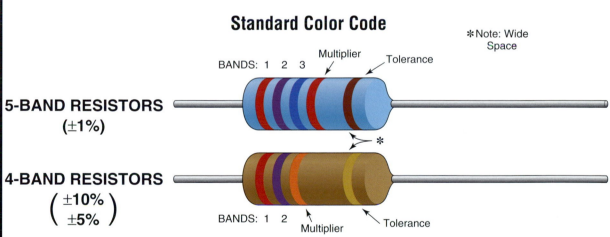

BANDS: 1 2 3 Multiplier Tolerance

5-BAND RESISTORS
(±1%)

4-BAND RESISTORS
$\left(\begin{array}{c}\pm\textbf{10\%} \\ \pm\textbf{5\%}\end{array}\right)$

BANDS: 1 2 Multiplier Tolerance

Band 1 First Digit		Band 2 Second Digit		Band 3 (if used) Third Digit		Multiplier		Resistance Tolerance	
Color	Digit	Color	Digit	Color	Digit	Color	Multiplier	Color	Tolerance
Black	0	Black	0	Black	0	Black	1	Silver	±10%
Brown	1	Brown	1	Brown	1	Brown	10	Gold	± 5%
Red	2	Red	2	Red	2	Red	100	Brown	± 1%
Orange	3	Orange	3	Orange	3	Orange	1,000		
Yellow	4	Yellow	4	Yellow	4	Yellow	10,000		
Green	5	Green	5	Green	5	Green	100,000		
Blue	6	Blue	6	Blue	6	Blue	1,000,000		
Violet	7	Violet	7	Violet	7	Silver	0.01		
Gray	8	Gray	8	Gray	8	Gold	0.1		
White	9	White	9	White	9				

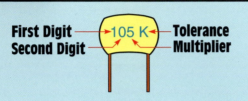

First Digit — 105 K — **Tolerance**
Second Digit — **Multiplier**

NUMBER	MULTIPLIER		TOLERANCE	
			10 pF or less	Over 10 pF
0	1	B	0.1 pF	
1	10	C	0.25 pF	
2	100	D	−0.5 pF	
3	1,000	F	1.0 pF	1%
4	10,000	G	2.0 pF	2%
5	100,000	H		3%
6		J		5%
7		K		10%
8	0.01	M		20%
9	0.1			

Film-type capacitors

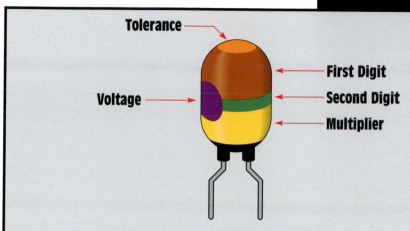

COLOR	NUMBER	MULTIPLIER	TOLERANCE (%)	VOLTAGE
			No dot 20	
Black	0			4
Brown	1			6
Red	2			10
Orange	3			15
Yellow	4	10,000		20
Green	5	100,000		25
Blue	6	1,000,000		35
Violet	7	10,000,000		50
Gray	8			
White	9			3
Gold			5	
Silver			10	

Dipped tantalum capacitors

STANDARD RESISTANCE VALUES (Ω)

0.1% 0.25% 0.5%	1%	0.1% 0.25% 0.5%	1%	0.1% 0.25% 0.5%	1%	0.1% 0.25% 0.5%	1%	0.1% 0.25% 0.5%	1%
10.0	10.0	17.2	–	29.4	29.4	50.5	–	86.6	86.6
10.1	–	17.4	17.4	29.8	–	51.1	51.1	87.6	–
10.2	10.2	17.6	–	30.1	30.1	51.7	–	88.7	88.7
10.4	–	17.8	17.8	30.5	–	52.3	52.3	89.8	–
10.5	10.5	18.0	–	30.9	30.9	53.0	–	90.9	90.9
10.6	–	18.2	18.2	31.2	–	53.6	53.6	92.0	–
10.7	10.7	18.4	–	31.6	31.6	54.2	–	93.1	93.1
10.9	–	18.7	18.7	32.0	–	54.9	54.9	94.2	–
11.0	11.0	18.9	–	32.4	32.4	55.6	–	95.3	95.3
11.1	–	19.1	19.1	32.8	–	56.2	56.2	96.5	–
11.3	11.3	19.3	–	33.2	33.2	56.9	–	97.6	97.6
11.4	–	19.6	19.6	33.6	–	57.6	57.6	98.8	–
11.5	11.5	19.8	–	34.0	34.0	58.3	–		
11.7	–	20.0	20.0	34.4	–	59.0	59.0		
11.8	11.8	20.3	–	34.8	34.8	59.7	–		
12.0	–	20.5	20.5	35.2	–	60.4	60.4		
12.1	12.1	20.8	–	35.7	35.7	61.2	–		
12.3	–	21.0	21.0	36.1	–	61.9	61.9		
12.4	12.4	21.3	–	36.5	36.5	62.6	–		
12.6	–	21.5	21.5	37.0	–	63.4	63.4		
12.7	12.7	21.8	–	37.4	37.4	64.2	–	2%, 5%	10%
12.9	–	22.1	22.1	37.9	–	64.9	64.9	10	10
13.0	13.0	22.3	–	38.3	38.3	65.7	–	11	–
13.2	–	22.6	22.6	38.8	–	66.5	66.5	12	12
13.3	13.3	22.9	–	39.2	39.2	67.3	–	13	–
13.5	–	23.2	23.2	39.7	–	68.1	68.1	15	15
13.7	13.7	23.4	–	40.2	40.2	69.0	–	16	–
13.8	–	23.7	23.7	40.7	–	69.8	69.8	18	18
14.0	14.0	24.0	–	41.2	41.2	70.6	–	20	–
14.2	–	24.3	24.3	41.7	–	71.5	71.5	22	22
14.3	14.3	24.6	–	42.2	42.2	72.3	–	24	–
14.5	–	24.9	24.9	42.7	–	73.2	73.2	27	27
14.7	14.7	25.2	–	43.2	43.2	74.1	–	30	–
14.9	–	25.5	25.5	43.7	–	75.0	75.0	33	33
15.0	15.0	25.8	–	44.2	44.2	75.9	–	36	–
15.2	–	26.1	26.1	44.8	–	76.8	76.8	39	39
15.4	15.4	26.4	–	45.3	45.3	77.7	–	43	–
15.6	–	26.7	26.7	45.9	–	78.7	78.7	47	47
15.8	15.8	27.1	–	46.4	46.4	79.6	–	51	–
16.0	–	27.4	27.4	47.0	–	80.6	80.6	56	56
16.2	16.2	27.7	–	47.5	47.5	81.6	–	62	–
16.4	–	28.0	28.0	48.1	–	82.5	82.5	68	68
16.5	16.5	28.4	–	48.7	48.7	83.5	–	75	–
16.7	–	28.7	28.7	49.3	–	84.5	84.5	82	82
16.9	16.9	29.1	–	49.9	49.9	85.6	–	91	–

Standard resistance values

Q_6 to stator winding T_3 through the motor to T_2 and through Q_3 to the positive of the power supply.

Because transistors are turned completely on or completely off, the waveform produced is a square wave instead of a sine wave, Figure 28-4. Induction motors will operate on a square wave without a great deal of problem. Some manufacturers design units that will produce a stepped waveform, as shown in Figure 28-5. The stepped waveform is used because it closer approximates a sine wave.

Some Related Problems

The circuit illustrated in Figure 28-2 employs the use of SCRs in the power supply and junction transistors in the output stage. SCR power supplies control the output voltage by chopping the incoming waveform. This can cause harmonics on the line that cause overheating of transformers and motors, and can cause fuses to blow and circuit breakers to trip. When bipolar junction transistors are employed as switches, they are generally driven into saturation by supplying an excessive amount of base-emitter current. Saturating the transistor causes the collector–emitter voltage to drop to between 0.04 and 0.03 volts. This small drop allows the transistor to control large amounts of current without being destroyed. When a transistor is driven into saturation, however, it cannot recover or turn off as quickly as normal. This greatly limits the frequency response of the transistor.

IGBTs

Many transistor-controlled variable drives now employ a special type of transistor called an insulated gate bipolar transistor (IGBT). IGBTs have an insulated gate very similar to some types of field effect transistors (FETs). Because the gate is insulated, it has a very high impedance. The IGBT is a voltage-controlled device, not a current-controlled device. This gives it the ability to turn off very quickly. IGBTs can be driven into saturation to provide a very low voltage drop between emitter and collector, but they do not suffer from the slow recovery time of common junction transistors. The schematic symbol for an IGBT is shown in Figure 28-6.

Drives using IGBTs generally used diodes to rectify the AC voltage into DC, not SCR, Figure 28-7. The three-phase rectifier supplies a constant DC voltage to the transistors. The output voltage to the motor is controlled by pulse width modulation (PWM). PWM is accomplished by turning the transistor on and off several times during each half cycle. The output voltage is an average of the peak or maximum voltage and the amount of time the transistor is turned on or off. Assume that 480 volts of three-phase AC is rectified to DC and filtered. The DC voltage applied to the IGBTs is approximately

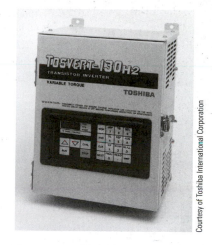

FIGURE 28-3 *Variable frequency drive using transistors in the output stage*

Courtesy of Toshiba International Corporation

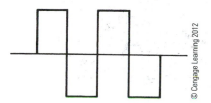

FIGURE 28-4 *Square-wave AC*

© Cengage Learning 2012

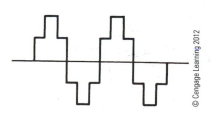

FIGURE 28-5 *Stepped square wave*

© Cengage Learning 2012

FIGURE 28-6 *Schematic symbol for an insulated gate bipolar transistor (IGBT)*

© Cengage Learning 2012

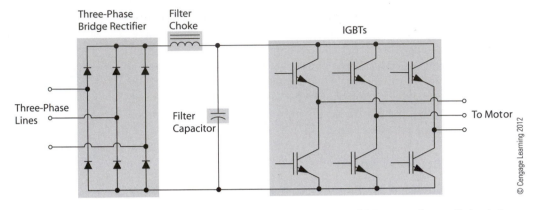

FIGURE 28-7 *Variable frequency drives using IGBTs generally use diodes in the rectifier instead of SCRs.*

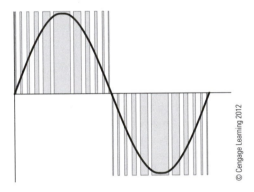

FIGURE 28-8 *The speed of the IGBTs can produce a stepped wave that is similar to a sine wave.*

630 volts. The output voltage to the motor is controlled by switching of the transistors. Assume that the transistor is on for 10 μs and off for 20 μs. In this example the transistor is on for one-third of the time and off for two-thirds of the time. The voltage applied to the motor would be 210 volts (630/3). By varying the on and off times during each half cycle, the output voltage can closely approximate a sine wave, Figure 28-8.

Advantages and Disadvantages of IGBT Drives

A great advantage of drives using IGBTs is the fact that SCRs are generally not used in the power supply and this greatly reduces problems with line harmonics. The greatest disadvantage is that the fast switching rate of the transistors can cause voltage spikes in the range of 1600 volts to be applied to the motor. These voltage spikes can destroy some motors. Line length from the drive to the motor is of great concern with drives using IGBTs. The shorter the line length, the better.

Inverter Rated Motors

Due to the problem of excessive voltage spikes caused by IGBT drives, some manufacturers produce a motor that is "inverter rated." These motors are specifically designed to be operated by variable frequency drives. They differ from standard motors in several ways:

1. Many inverter rated motors contain a separate blower to provide continuous cooling for the motor regardless of the speed. Many motors use a fan connected to the motor shaft to help draw air through the motor. When the motor speed is reduced, the fan cannot maintain sufficient air flow to cool the motor.

2. Inverter rated motors generally have insulating paper between the windings and the stator core, Figure 28-9. The high voltage spikes produce high currents that produce a high magnetic field. This increased magnetic field causes the motor windings to move. This movement can eventually cause the insulation to wear off the wire and produce a grounded motor winding.

3. Inverter rated motors generally have phase paper added to the terminal leads. Phase paper is insulating paper added to the terminal leads that exit the motor. The high-voltage spikes affect the beginning lead of a coil much more than the wire inside the coil. The coil is an inductor that naturally opposes a change of current. Most of the insulation stress caused by high-voltage spikes occurs at the beginning of a winding.

4. The magnet wire used in the construction of the motor windings has a higher rated insulation than other motors.

5. The case size is larger than most three-phase motors. The case size is larger because of the added insulating paper between the windings and the stator core. Also, a larger case size helps cool the motor by providing a larger surface area for the dissipation of heat.

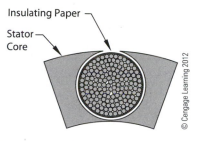

FIGURE 28-9 *Insulating paper is between the windings and the stator frame.*

© Cengage Learning 2012

Variable Frequency Drives Using SCRs and GTOs

Variable frequency drives intended to control motors over 500 Hp generally use SCRs or GTOs (gate turn-off devices). GTOs are similar to SCRs except that conduction through the GTO can be stopped by applying a negative voltage, negative with respect to the cathode, to the gate. SCRs and GTOs are thyristors and have the ability to handle a greater amount of current than transistors. An example of a single-phase circuit used to convert DC voltage to

FIGURE 28-10 *Changing DC into AC using SCRs*

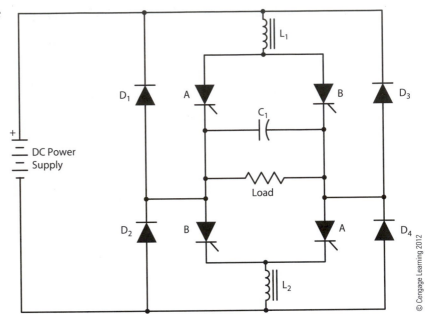

© Cengage Learning 2012

AC voltage with SCRs is shown in Figure 28-10. In this circuit, the SCRs are connected to a control unit that controls the sequence and rate at which the SCRs are gated on. The circuit is constructed so that SCRs A and A′ are gated on at the same time and SCRs B and B′ are gated on at the same time. Inductors L_1 and L_2 are used for filtering and wave shaping. Diodes D_1 through D_4 are clamping diodes and are used to prevent the output voltage from becoming excessive. Capacitor C_1 is used to turn one set of SCRs off when the other set is gated on. This capacitor must be a true AC capacitor because it will be charged to the alternate polarity each half cycle. In a converter intended to handle large amounts of power, capacitor C_1 will be a bank of capacitors. To understand the operation of the circuit, assume that SCRs A and A′ are gated on at the same time. Current will flow through the circuit as shown in Figure 28-11. Notice the direction of current flow through the load, and that capacitor C_1 has been charged to the polarity shown. When an SCR is gated on, it can only be turned off by permitting the current flow through the anode–cathode section to drop below a certain level called the holding current level. As long as the current continues to flow through the anode–cathode, the SCR will not turn off.

Now assume that SCRs B and B′ are turned on. Because SCRs A and A′ are still turned on, two current paths now exist through the circuit. The positive charge on capacitor C_1, however, causes the negative electrons to see an easier path. The current will rush to charge the capacitor to the opposite polarity, stopping the current

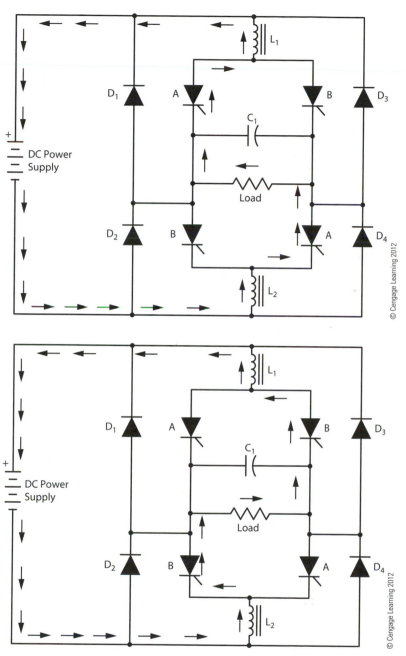

FIGURE 28-11 *Current flows through SCRs A and A'*

FIGURE 28-12 *Current flows through SCRs B and B'*

© Cengage Learning 2012

flowing through SCRs A and A', permitting them to turn off. The current now flows through SCRs B and B' and charges the capacitor to the opposite polarity, Figure 28-12. Notice that the current now flows through the load in the opposite direction, which produces alternating current across the load.

To produce the next half cycle of AC current, SCRs A and A′ are gated on again. The positively charged side of the capacitor will now cause the current to stop flowing through SCRs B and B′, permitting them to turn off. The current again flows through the load in the direction indicated in Figure 28-11. The frequency of the circuit is determined by the rate at which the SCRs are gated on.

Features of Variable Frequency Control

Although the primary purpose of a variable frequency drive is to provide speed control for an AC motor, most drives provide functions that other types of controls do not. Many variable frequency drives can provide the low-speed torque characteristic that is so desirable in DC motors. It is this feature that permits AC squirrel cage motors to replace DC motors for many applications.

Many variable frequency drives also provide current limit and automatic speed regulation for the motor. Current limit is generally accomplished by connecting current transformers to the input of the drive and sensing the increase in current as load is added. Speed regulation is accomplished by sensing the speed of the motor and feeding this information back to the drive.

Another feature of variable frequency drives is acceleration and deceleration control, sometimes called "ramping." Ramping is used to accelerate or decelerate a motor over some period of time. Ramping permits the motor to bring the load up to speed slowly as opposed to simply connecting the motor directly to the line. Even if the speed control is set in the maximum position when the start button is pressed, ramping permits the motor to accelerate the load from zero to its maximum RPM over several seconds. This feature can be a real advantage for some types of loads, especially gear drive loads. In some units, the amount of acceleration and deceleration time can be adjusted by setting potentiometers on the main control board. Other units are completely digitally controlled and the acceleration and deceleration times are programmed into the computer memory.

Some other adjustments that can usually be set by changing potentiometers or programming the unit are as follows:

Current Limit: This control sets the maximum amount of current the drive is permitted to deliver to the motor.

Volts per Hertz: This sets the ratio by which the voltage increases as frequency increases or decreases as frequency decreases.

Maximum Hz: This control sets the maximum speed of the motor.

Minimum Hz: This sets the minimum speed at which the motor is permitted to run.

Some variable frequency drives permit adjustment of current limit, maximum and minimum speed, ramping time, and so on, by adjustment of trim resistors located on the main control board. Other drives employ a microprocessor as the controller. The values of current limit, speed, ramping time, and so forth, for these drives are programmed into the unit and are much easier to make and generally more accurate than adjusting trim resistors.

CONVERTING FROM DC TO AC

The oscillator shown in Figure 28-13 is much simpler than the ones just described. It is used to convert direct current into single-phase alternating current. The output voltage is a square wave. The frequency is determined by the turns of the transformer and the applied voltage.

This circuit will operate because of the change of the magnetic structure in the core of the transformer. When the power is first applied, one transistor will begin to conduct before the other. This is due to slight imbalances in the characteristics of the transistors. It will be assumed that transistor Q_1 will conduct first. When transistor Q_1 turns on, current will flow from point B to point A, through Q_1 and back to the negative of the power supply. As current flows through the transformer winding, a magnetic field is produced, which causes the iron molecules in the core to align themselves in one direction. When all the molecules have become aligned, the transformer core becomes saturated. This causes the voltage between points B and A to drop to a very low value, causing transistor Q_1 to turn off.

Transistor Q_2 now turns on and permits current to flow from point B to point C and through transistor Q_2 to the negative terminal of the power supply. Because current is now flowing through the

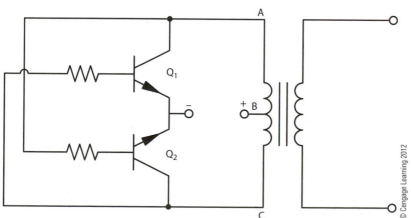

FIGURE 28-13 *Single-phase square-wave oscillator*

© Cengage Learning 2012

winding in an opposite direction, the magnetic field reverses polarity. This causes the molecules of iron in the transformer core to realign themselves in the opposite direction. When they have all become aligned, the transformer core again becomes saturated and the voltage between points B and C drops to a low value, causing transistor Q_2 to turn off. Transistor Q_1 again turns on and the process starts again.

Each time current flows through the primary of the transformer, it induces a voltage into the secondary. Because the current flows first in one direction and then in the other, the induced voltage is AC. The transistors in this circuit are either completely on or completely off. The output voltage is square-wave alternating current.

UNIT 28 REVIEW QUESTIONS

1. What is an oscillator? _____

_____

2. When an oscillator is used to control the synchronous field speed of an AC induction motor by
lowering the frequency, the applied voltage to the motor must be reduced. Why must the voltage be
reduced as the frequency is lowered? _____

_____

_____

3. Why do some oscillators produce a stepped waveform? _____

_____

4. What type of waveform is produced by the oscillator in Figure 28-13? _____

_____

5. What determines the output frequency of the oscillator in Figure 28-13? _____

_____

UNIT 29

THE DC TO DC VOLTAGE DOUBLER

OBJECTIVES

After studying this unit, the student should be able to:

■ Discuss the operation of a voltage doubler.

■ Construct a voltage doubler using discrete electronic components.

There may be occasions when it is necessary or desirable to have a higher DC voltage than what is available. If a higher AC voltage were needed, it would be a simple matter. A transformer with the proper turns ratio and power rating could be found and connected into the circuit. Direct current, however, cannot be transformed. It must first be converted into alternating current by the use of an oscillator. It is then raised to a higher voltage and rectified back to direct current.

In the previous unit, an oscillator was constructed like the one shown in Figure 29-1. The same basic circuit can be used to construct a circuit that will double the DC voltage applied to it. In the circuit in Figure 29-2, the transformer shown is a single winding. It has been center tapped, similar to a center-tapped auto transformer. The collector of each transistor is connected to the anode of a diode.

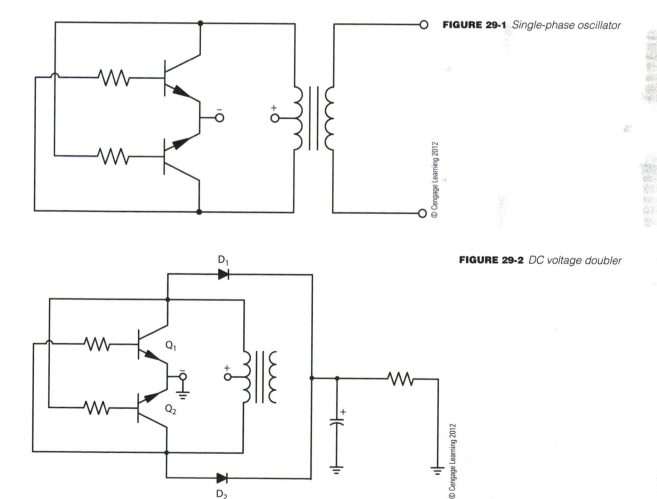

FIGURE 29-1 *Single-phase oscillator*

FIGURE 29-2 *DC voltage doubler*

© Cengage Learning 2012

The cathodes of the diodes are connected to form a two-diode full-wave rectifier. The output of the rectifier is connected to a capacitor filter and then to the load. The load is connected to ground, the negative of the DC input.

Assume that a voltage of 6 volts DC is applied to the oscillator. When transistor Q_1 turns on, current flows through half of the winding of the transformer to point A. Diode D_1 is forward biased and conducts through the load to ground. Transistor Q_1 turns off and Q_1 turns on. Current flows through the other half of the winding to point C. Diode D_2 is now forward biased. Current flows through the diode to the load resistor and to ground.

Because 6 volts has been applied across each half of the winding, the total output voltage is 12 volts. The oscillator produces a square wave AC voltage that is double the DC voltage applied to it. The rectifier then changes the AC voltage back into direct current.

UNIT 29 REVIEW QUESTIONS

1. What must be done before direct current can be transformed to a higher voltage? _____

_____

2. In the circuit used in this experiment, what type of rectifier is used to change the AC voltage back into DC? _____

_____

3. What type of waveform is produced by the oscillator in Figure 29-2? _____

_____

4. What is the function of the 1000-μF capacitor used in Laboratory Exercise 27, Figure Exp. 27-1? __

_____

UNIT 30

THE OFF-DELAY TIMER

OBJECTIVES

After studying this unit, the student should be able to:

- Describe the operation of an off-delay timer.

- Draw the schematic symbol for an off-delay timer contact.

- Construct an off-delay timer circuit using discrete electronic components.

Timers have been used in industry since the beginning of individual-ized motor controls. Among these are dashpot, pneumatic, clock, and electronic timers. Regardless of the method used to achieve a time delay for the relay, timers have only one function. They delay chang-ing their contact position when they are turned on or off.

On-delay timers delay when they are turned on, but change back immediately when turned off. They are often referred to as DOE on a schematic. This means "delay on energize."

Off-delay timers, however, change their contacts immediately when turned on. They delay changing back, however, when turned off. Relay contacts are always shown on a schematic in their de-energize (off) position.

BASICS OF OPERATION

Refer to Figure 30-1, which shows the NEA symbol for an off-delay, normally open contact. When the timer is energized, the contact closes immediately just as any normal relay contact will. The contact remains closed as long as power is applied to the relay. When the relay is de-energized (turned off), the contact does not reopen immediately like a common relay contact. It remains closed for a predetermined amount of time before it reopens. Off-delay relays are often referred to on schematics as DODE. This means "delay on de-energize."

Methods of Control

There are various methods used to obtain the time delay needed. This text is primarily concerned with electronic methods. There are several different ways of obtaining time delays using electronic components, but only two will be discussed. The simplest method is shown in Figure 30-2.

SIMPLE TIMER: K_1 is the coil of a 12-volt DC relay, and C_1 is a capacitor connected in parallel with the coil. When switch S_1 is closed, K_1 immediately energizes. Capacitor C_1 is charged to 12 volts DC. When switch S_1 is opened, capacitor C_1 begins to discharge through K_1. This keeps K_1 turned on as long as C_1 can supply enough current to hold the relay in. This circuit uses a simple RC time constant. The amount of time K_1 remains energized is determined by the resistance of coil K_1 and the capacitance of C_1. The circuit is useful when the

Normally Open
Off-delay Contact

Normally Closed
Off-delay Contact

FIGURE 30-1 *Off-delay contact symbols.*

© Cengage Learning 2012

FIGURE 30-2 *Simple capacitor timer*

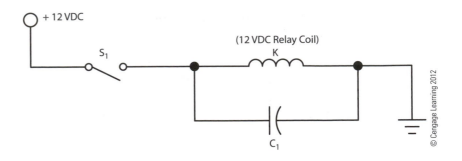

time delay needed is a few seconds or less. If longer time delays are needed, the value of C_1 would become very high.

ELECTRONIC TIMER: The more complex circuit shown in Figure 30-3 can be used for delay times of a few seconds to an hour or more if desired. The circuit is basically a Darlington amplifier circuit. It operates as follows:

- Transistor Q_1 is connected in series with the relay coil K_1. It is used to turn K_1 on or off.

FIGURE 30-3 *An electronic off-delay timer*

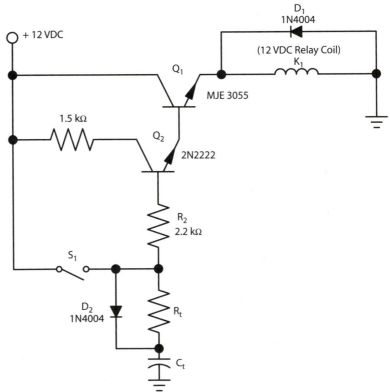

- Transistor Q_1 amplifies the base of transistor Q_1.

- Resistor R_1 limits the current to the base of Q_1.

- Resistor R_2 limits the current to the base of Q_1 when switch S_1 is closed.

- Switch S_1 controls the operation of the circuit. When it is closed, relay K_1 immediately turns on. When it is opened, relay K_1 is delayed in turning off by the RC time constant of R_t and C_t.

- Diode D_1 is known as a kick back or free-wheeling diode. Its job is to kill the spike voltage created when the current through K_1 is suddenly stopped. (Notice that diode D_1 is connected in the circuit in the reverse bias position.)

  1. When current flows through K_1, a magnetic field is developed around the coil.

  2. If the current is suddenly interrupted, the collapsing magnetic field induces a high voltage in the coil. The induced voltage can be a spike of several hundred volts. A spike voltage this large can destroy electronic components throughout the circuit.

  3. Diode D_1 is reverse biased to the applied voltage of the circuit. (Note: An induced voltage is always opposite in polarity to the applied voltage.)

  4. That makes D_1 forward biased to any spike voltages induced in the coil of K_1.

  5. The forward voltage drop of a silicon diode is only about 0.7 volt.

  6. Diode D_1 will not permit the induced voltage of K_1 to become greater than its forward voltage drop.

- Diode D_1 is used to bypass resistor R_t so capacitor D_1 can be charged immediately when switch S_1 is closed. If diode D_1 were removed from the circuit, capacitor C_t would have to be charged through resistor R_t. This could take several minutes, depending on the values of R_t and C_t. When switch S_1 is opened, the charge of capacitor C_t makes diode D_1 reverse biased. C_t therefore discharges through resistor R_t.

- Capacitor C_t supplies current to the base of transistor Q_2 after S_1 has been opened.

- Relay K_1 remains turned on as long as capacitor C_1 can supply enough current to the base of Q_2 to keep the circuit turned on.

- Resistor R_t determines the discharge time of C_t. The values of C_t and R_t determine the delay time of the circuit.

UNIT 30 REVIEW QUESTIONS

1. When the coil of an off-delay timer is energized, do the contacts change position immediately or do they delay changing positions? _____

   _____

2. When the coil of an off-delay timer is de-energized, do the contacts change back immediately or do they delay changing back to their original off position? _____

   _____

3. In the timer circuit shown in the Figure 30-2, what factors determine the amount of delay time?

   _____

   _____

4. What does DODE stand for? _____

   _____

5. What is the function of diode D_1 in the schematic shown in Figure 30-3? _____

   _____

   _____

6. What is the function of diode D_2 in the schematic shown in Figure 30-3? _____

   _____

UNIT 31

THE ON-DELAY TIMER

OBJECTIVES

After studying this unit, the student should be able to:

- Discuss the operation of an on-delay timer.

- Draw the schematic symbol for an on-delay timer contact.

- Construct an on-delay timer using discrete electronic components.

The on-delay timer is used throughout industry in many applications. The time delay can be accomplished by several methods: pneumatic, electric clock, or electronic. A quartz clock is used in some electronic timers that must be very accurate. Most electronic timers, however, use an RC time constant to do the job. They are inexpensive and relatively accurate. A variable resistor added to the circuit allows adjustment of the length of time delay within the limits of the circuit.

Regardless of the method used, all on-delay timers operate basically the same way. When an on-delay relay is energized, its contacts remain in their normal position for some length of time before they change. When the relay is de-energized, the contacts return to their normal position immediately.

Figure 31-1 shows the NEMA (National Electrical Manufacturers Association) symbol for a normally open set of contacts operated by an on-delay relay. Assume that this contact is controlled by an on-delay relay that has been set for 10 seconds. When the relay is energized, the contact remains open for 10 seconds and then closes. When the relay is de-energized, the contact returns to its open position immediately.

The circuit shown in Figure 31-2 is a simple on-delay time circuit. It uses a unijunction transistor as the timing element. The values of R_t and C_1 determine the timing of the circuit. Resistor R_1 limits the current through the UJT. R_2 permits a positive pulse to be produced across the resistor when the UJT turns on and discharges

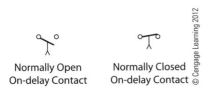

Normally Open
On-delay Contact

Normally Closed
On-delay Contact

© Cengage Learning 2012

FIGURE 31-1 *On-delay contact symbols*

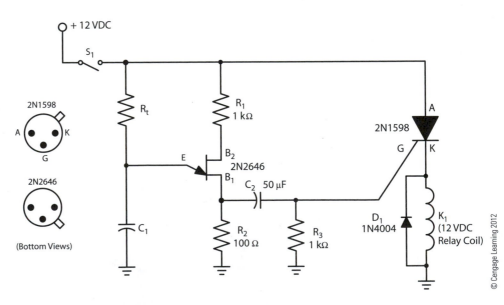

© Cengage Learning 2012

FIGURE 31-2 *An on-delay timer circuit*

capacitor C_1. If resistor R_2 were removed and the circuit grounded, a positive pulse could not be produced when C_1 is discharged.

Capacitor C_2 isolates the gate of the SCR from the UJT. Any leakage current through the UJT is blocked by C_2. A large pulse caused by the discharge of C_1 is passed across C_2 like an AC voltage would be.

Resistor R_3 keeps the gate of the SCR at a ground potential until a pulse is transmitted through capacitor C_2 to fire the gate of the SCR. When the SCR fires, it provides current to K_1; the coil of a 12-volt DC relay coil. Once the SCR fires, it remains turned on until the circuit is broken by switch S_1. Diode D_1 is used as a kick back or free-wheeling diode. It kills the voltage spike induced in the coil of K_1 when the circuit is opened.

Note: Times will vary with different components due to the difference in gain. Some of these time delays were made extremely long to demonstrate the flexibility of this timer circuit.

UNIT 31 REVIEW QUESTIONS

1. When the coil of an on-delay timer is energized, do the contacts change position immediately or do they delay changing position for some period of time? _____

_____

2. When the coil of an on-delay timer is de-energized, do the contacts return to their normal position immediately, or do they delay changing back? _____

_____

3. In the circuit shown in Figure 31-2, what determines the amount of time delay? _____

_____

4. What is the function of resistor R_2 in the circuit shown in Figure 31-2? _____

_____

5. Once the SCR in Figure 31-2 has been turned on, what must be done to turn it off? _____

_____

UNIT 32

THE PULSE TIMER

OBJECTIVES

After studying this unit, the student should be able to:

- Discuss the operation of a pulse timer.

- Construct a pulse timer using discrete electronic components.

The pulse timer is similar in construction to the on-delay timer, with one major difference. When the on-delay timer is turned on, it remains that way until the power is interrupted. The pulse timer, however, turns itself back off after it has been turned on. The windshield wiper delay circuit on many automobiles is a good example of this type of circuit.

The circuit shown in Figure 32-1 is a simple pulse-timer circuit. The unijunction transistor, Q_1, is used to provide a pulse. The time between pulses is determined by the RC time constant of $R_1 + R_2$ and C_1. When the voltage across capacitor C_1 becomes high enough, the UJT turns on and discharges C_1 through resistor R_6 to ground. As the capacitor discharges through R_6, a positive voltage spike appears across R_6.

Transistor Q_3 is connected in series with K_1, which is the coil of a 12-volt DC relay. Resistor R_5 limits the base current for transistor Q_3. When transistor Q_3 turns on, coil K_1 also turns on. Capacitor C_3 is connected to the base of transistor Q_3. It is used to keep the base of transistor Q_3 at ground until a positive voltage appears at the base of Q_3.

Transistor Q_2 is used as a stealer transistor. A stealer transistor steals the base current from another transistor to keep it turned off. In this circuit, Q_2 steals the current from the base of transistor Q_3 to keep it turned off. As long as transistor Q_2 is turned on, this condition will continue. The resistance of $R_3 + R_4$ determines the amount of base current to transistor Q_2. Resistor R_4 ensures that the current to the base of Q_2 will be limited if the value of resistor R_3 is adjusted to 0 Ω.

FIGURE 32-1 *Electronic pulse timer*

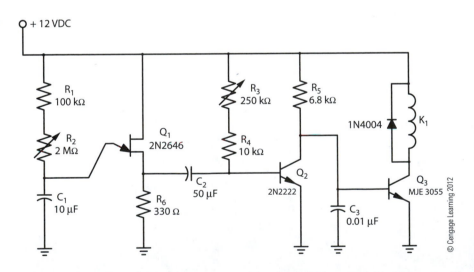

© Cengage Learning 2012

Capacitor C_2 is charged through the resistance of $R_3 + R_4$. When capacitor C_2 is charged sufficiently, base current keeps transistor Q_2 turned on. When transistor Q_2 is turned on, the base of transistor Q_3 is at ground potential and therefore off.

When the UJT turns on and discharges capacitor C_1, a large positive pulse is produced at capacitor C_2. This large positive pulse causes the base-emitter junction of transistor Q_2 to become forward biased. Electrons flow from ground through the base-emitter junction to the capacitor. Because electrons are negative particles, the base of transistor Q_2 becomes negative, which turns the transistor off. This permits base current to flow to transistor Q_3 and turn on relay K_1. Transistor Q_2 will remain turned off until the capacitor C_2 can again be charged positive through resistors R_3 and R_4. In actual circuit operation, resistor R_2 determines the time between pulses, and resistor R_3 determines the amount of time the relay stays turned on.

UNIT 32 REVIEW QUESTIONS

1. Explain the difference in operation between an on-delay timer and a pulse timer. _____
_____
_____

2. In the circuit shown in Figure 32-1, what is the function of the UJT? _____
_____

3. What is the function of resistor R_2 in Figure 32-1? _____
_____

4. What is the function of resistor R_3 in Figure 32-1? _____
_____

5. What is a stealer transistor? _____
_____

UNIT 33

THE 555 TIMER

OBJECTIVES

After studying this unit, the student should be able to:

- Discuss the operation of a 555 timer.

- List the pins on the timer and give an explanation of what each does.

- Connect the 555 timer in an electronic circuit.

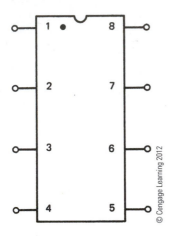

FIGURE 33-1 *Eight-pin integrated circuit*

PIN EXPLANATION

The 555 timer is an eight pin inline integrated circuit (IC), Figure 33-1. This package will have a notch at one end or a dot near one pin. These identify pin #1. Once pin #1 has been identified, the other pins are numbered as shown. The following is an explanation of each pin and what it does.

Pin #1   Ground—This pin is connected to circuit ground.

Pin #2   Trigger—This pin must be connected to a voltage that is less than one-third Vcc (the applied voltage) to trigger the unit. This is generally done by connecting pin #2 to ground. The connection to one-third Vcc or ground must be momentary. If pin #2 is not removed from ground, the unit will not operate.

Pin #3   Output—The output turns on when pin #2 is triggered, and off when the discharge is turned on.

Pin #4   Reset—When this pin is connected to Vcc, it permits the unit to operate. When connected to ground, it activates the discharge and keeps the timer from operating.

Pin #5   Control voltage—If this pin is connected to Vcc through a variable resistor, the on time becomes longer. The off time remains unaffected. If it is connected to ground through a variable resistor, the on time becomes shorter. The off time is still not affected. If it is not used in the circuit, it is generally taken to ground through a small capacitor. This prevents circuit noise from talking to pin #5.

Pin #6   Threshold—When the voltage across the capacitor connected to pin #6 reaches two-thirds of the value of Vcc, the discharge turns on and the output turns off.

Pin #7   Discharge—When pin #6 turns the discharge on, it discharges the capacitor connected to pin #6. The discharge remains turned on until pin #2 retriggers the timer. It then turns off and the capacitor connected to pin #6 begins charging again.

Pin #8   Vcc—This pin is connected to the applied voltage of the circuit that is known as Vcc. The 555 timer operates on a wide range of voltages. Its operating voltage range is considered to be between 3 and 16 volts DC.

(Note: For the following explanation, assume that pin #2 is connected to pin #6. This permits the unit to be retriggered by the discharge each time it turns on and discharges the capacitor to one-third the value of Vcc.)

OPERATION OF THE TIMER

The 555 timer operates on a percentage of the applied voltage. The time setting remains constant even if the applied voltage changes. When the capacitor connected to pin #6 reaches two-thirds of the applied voltage, the discharge turns on. The capacitor discharges until it reaches one-third of the applied voltage. If Vcc of the timer is connected to 12 volts DC, two-thirds of the applied voltage is 8 volts and one-third is 4 volts. This means that when the voltage across the capacitor connected to pin #6 reaches 8 volts, pin #7 turns on. After the capacitor discharges to one-third the value of Vcc, or 4 volts, it turns off, Figure 33-2.

If the voltage is lowered to 6 volts at Vcc, two-thirds of the applied voltage becomes 4 volts and one-third is 2 volts. Pin #7 now turns on when the voltage across the capacitor connected to pin #6 reaches 4 volts. It turns off when the voltage across the capacitor drops to 2 volts. The formula for an RC time constant is time = resistance × capacitance. There is no mention of voltage in the formula. It takes just as long to charge the capacitor if the circuit is connected to 12 volts as it will for 6 volts. The voltage of the capacitor connected to pin #6 reaches two-thirds of Vcc the same way. When the timer has an applied voltage of 12 volts, it takes the same amount of time as when the applied voltage is only 6 volts. Notice that the timing of the circuit remains the same even if the voltage changes.

Using a Circuit to Explain the Timer

The circuit shown in Figure 33-3 is used to help explain the operation of the 555 timer. The figure shows a normally closed switch, S_1, connected between the discharge, pin #7, and the ground pin #1. There is a normally open switch, S_2, connected between the output, pin #3, and Vcc at pin #8. The dotted line drawn between these two switches shows mechanical connection: both switches operate together. If S_1 opens, S_2 closes at the same time. If S_2 opens, S_1 closes. Pin #2, the trigger, and pin #6, the threshold, are used to control these two switches. The trigger can close switch S_2, and the threshold can close S_1. To begin this analysis, assume switch S_1 is closed and switch S_2 is open, as shown in Figure 33-3.

1. When the trigger is connected to a voltage less than one-third of Vcc, switch S_2 closes and switch S_1 opens.

2. When switch S_2 closes, voltage is supplied to the output at pin #3.

3. When switch S_1 opens, the discharge is no longer connected to ground.

4. Capacitor C_1 begins to charge through resistor R_1 and R_2.

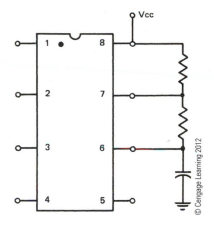

FIGURE 33-2 *RC time constant controls the operation of the timer.*

© Cengage Learning 2012

FIGURE 33-3 *Basic timer operation*

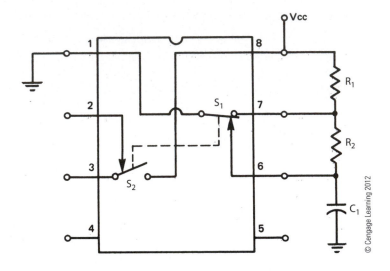

5. When the voltage across C_1 reaches two-thirds of Vcc, the threshold, pin #6, causes switch S_1 to close and switch S_2 to open.

6. When switch S_2 opens, the output turns off.

7. When switch S_1 closes, the discharge, pin #7, is connected to ground.

8. This permits capacitor C_1 to begin discharging through resistor R_2.

9. The state of the timer will remain in this position until the trigger is again connected to a voltage that is less than one-third of Vcc.

If the trigger is permanently connected to a voltage less than one-third of Vcc, switch S_2 is held closed and switch S_1 is held open. This, of course, stops the operation of the timer. The trigger must be a momentary pulse and not a continuous connection if the 555 is to operate.

APPLICATIONS

Thousands of uses for the 555 timer have been found in both commercial and industrial electronics. One application is shown in Figure 33-4. In this circuit, the 555 timer is the heart of a continuity tester that gives both a visual and an audible indication of when a complete circuit exists. When the test probes are connected together, visual proof of a complete circuit is provided by the LED, D_1. Resistor R_1 limits current flow through the LED.

The audible part of the tester is provided by the 555 timer and the small speaker. When the test probes are connected together, power is provided to the timer. Because pin #2 has been connected directly to pin #6, the timer will operate in an astable mode and provide a pulsating output at pin #3. Capacitor C_2 changes the pulsating

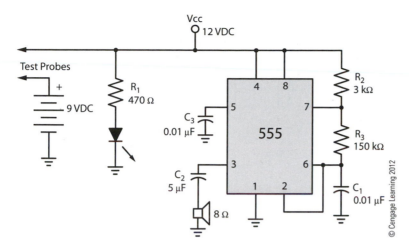

© Cengage Learning 2012

FIGURE 33-4 *Continuity tester*

DC into AC to operate the speaker. Resistor R_2 limits the current flow when the discharge pin, #7, turns on. The tone of the speaker can be adjusted by changing the value of resistor R_3 or capacitor C_1.

Another example of how a 555 timer can be used is shown in Figure 33-5. In this circuit, the timer is used to provide speed control to a direct current motor. The controller works by changing the amount of voltage applied to the armature of the motor. The 555 timer provides a pulsating DC voltage to the armature of the motor. The voltage applied across the armature is the average value determined by the length of time transistor Q_2 is turned on as compared to the length of time it is turned off. For example, the waveform in Figure 33-6 has a peak value of 12 volts. Notice the voltage is off three times longer than it is on. Assume the interval between pulsations to be 40 ms. If the voltage applied to the armature is turned on for 10 ms

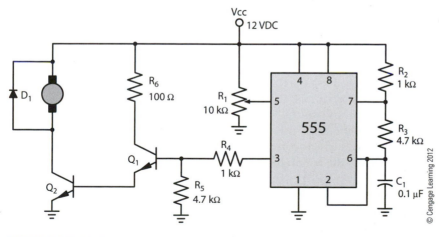

© Cengage Learning 2012

FIGURE 33-5 *DC motor speed control*

FIGURE 33-6 *Average DC voltage is 3 volts.*

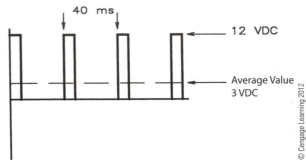

FIGURE 33-7 *Average voltage is 8 volts.*

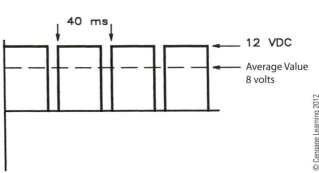

and turned off for 30 ms, the average voltage applied to the armature will be one-fourth the value of the total voltage, or 3 volts.

The waveform shown in Figure 33-7 indicates that the voltage is turned on for 30 ms and turned off for 10 ms. The average amount of voltage applied to the armature will now be three-fourths the total voltage, or 8 volts.

The 555 timer is used to control the amount of time transistor Q_2 is turned on or off. The timer is operated in the astable mode by connecting pin #2 to pin #6. Resistor R_2 limits the amount of current flow when the discharge turns on. Resistor R_3 and capacitor C_1 determine the amount of off time for the timer. Resistor R_4 limits current flow to the base of transistor Q_1, which is used as a Darlington driver for transistor Q_2. Resistor R_5 ensures transistor Q_1 remains turned off when there is no output at pin #3 of the timer. Resistor R_6 limits base current to transistor Q_2 when transistor Q_1 turns on. Diode D_1 performs the function of a commutating diode to prevent voltage spikes being produced when current flow through the armature stops. Resistor R_1 controls the length of time the output of the timer will be turned on, which controls the speed of the motor. If the wiper of resistor R_1 is adjusted close to Vcc, the output will be turned on for a long period of time as compared with the amount of time it will be turned off. If the wiper is adjusted close to ground, the output will be turned on for only a short period of time as compared with the off period.

UNIT 33 REVIEW QUESTIONS

1. What pin is used to ground the timer? _____

_____

2. Must the trigger pin be connected to a voltage greater than one-third of Vcc or a voltage less than one-third of Vcc for the timer to operate? _____

_____

3. What range of voltage can be used to operate the 555 timer? _____

_____

4. Does raising or lowering the input voltage affect the timing operation of the 555 timer? _____

_____

5. What happens when the reset pin is connected to ground? _____

_____

6. When two-thirds of Vcc is applied to the threshold, it turns on the discharge. What is used to turn off the discharge again? _____

_____

7. Does the output turn on or off when the timer is triggered? _____

_____

UNIT 34

THE 555 TIMER USED
AS AN OSCILLATOR

OBJECTIVES

After studying this unit, the student should be able to:

- Discuss the operation of a 555 timer when it is used as an oscillator.

- Connect a 555 timer as an oscillator.

- Make measurements of frequency using an oscilloscope.

The 555 timer can perform a variety of functions. One of the most common functions is that of an oscillator. The 555 timer has become popular for this application because it is very easy to use, Figure 34-1.

OPERATION OF THE TIMER

The 555 timer shown in Figure 34-1 has pin #2 connected to pin #6. This permits the timer to retrigger itself at the end of each time cycle. When the voltage at Vcc is first turned on, capacitor C_1 discharges and has a voltage of 0 volts across it. Because pin #2 is connected to pin #6, and the voltage at that point is less than one-third of Vcc, the timer triggers. When the timer triggers, two things happen at the same time: the output turns on and the discharge turns off. When the discharge at pin #7 turns off, capacitor C_1 begins charging through resistors R_1 and R_2. The time it takes for capacitor C_1 to charge is determined by its capacitance and the combined resistance of $R_1 + R_2$. When capacitor C_1 is charged to a voltage that is two-thirds of Vcc, two things happen: the output turns off and the discharge at pin #7 turns on. When the discharge turns on, capacitor C_1 discharges through resistor R_2 to ground. The time it takes C_1 to discharge is determined by the capacitance of C_1 and the resistance of R_2. When capacitor C_1 has discharged to where its voltage is one-third of Vcc, the timer is retriggered by pin #2. When the timer is retriggered, the output again turns on

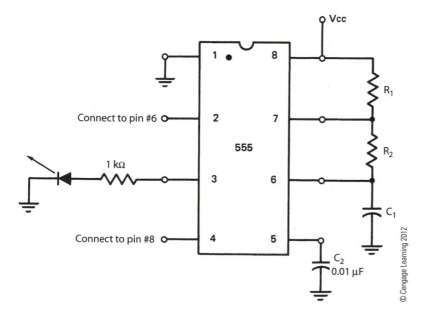

FIGURE 34-1 *A stable oscillator*

© Cengage Learning 2012

FIGURE 34-2 *The charge time is longer than the discharge time.*

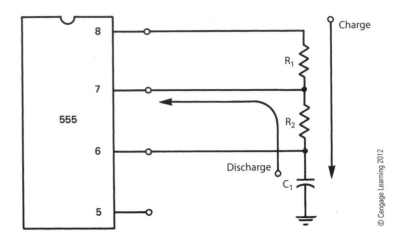

© Cengage Learning 2012

and the discharge turns off. When the discharge turns off, capacitor C_1 again begins to charge.

Note that the amount of time required to charge capacitor C_1 is determined by the combined resistance of $R_1 + R_2$. The discharge time, however, is determined by the value of R_2, Figure 34-2.

Discharge Time

The timer's output is turned on during the period when time capacitor C_1 is charging. It is turned off during the time C_1 is discharging. The on time of the output, therefore, is longer than the off time. If the value of resistor R_2 is much greater than resistor R_1, this condition is not too evident. For instance, if resistor R_1 has a value of 1 kilohm (kΩ) and R_2 has a value of 100 kΩ, the resistance connected in series with the capacitor during charging is 101 kΩ. The resistance connected in series with the capacitor during discharge is 100 kΩ. In this circuit, the charge time and the discharge time of the capacitor are 1 percent of each other. If an oscilloscope is connected to the output of the timer, a waveform similar to that shown in Figure 34-3 is seen.

Assume that the value of resistor R_1 is charged to 100 kΩ and that the value of resistor R_2 remains at 100 kΩ. In this circuit, the resistance connected in series with the capacitor during charging is 200 kΩ. The resistance connected in series with the capacitor during discharge, however, is 100 kΩ. In the circuit, the discharge time is 50 percent of the charge time. This means that the output of the timer is turned on twice as long as it is turned off. An oscilloscope

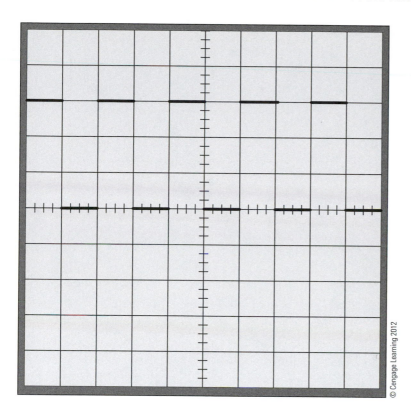

FIGURE 34-3 *On time and off time are the same.*

connected to the output of the timer displays a waveform similar to the one shown in Figure 34-4.

USING PIN #5: Although this condition can exist, the 555 timer has a provision for solving the problem. Pin #5, the control voltage pin, can give complete control of the on time of the timer, although it does not affect the off time. The off time is controlled by the amount of capacitance of C_1 and the resistance of R_2. If pin #5 is connected to Vcc through a resistor, the amount of on time will be greater and the off time will remain the same. If pin #5 is connected to ground through a resistor, the on time will be shorter and the off time will remain the same. Complete control of the on time can be gained by using a variable resistor as a potentiometer and connecting pin #5 to the wiper of the pot, as shown in Figure 34-5. In this circuit, the voltage applied to pin #5 can be varied between Vcc and ground, depending on the position of the wiper. This will give complete control of the on time of the timer.

FIGURE 34-4 *On time is longer than off time.*

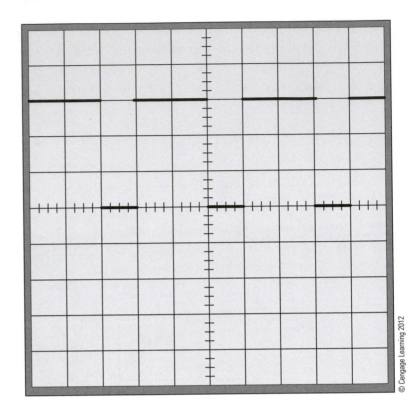

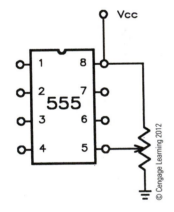

FIGURE 34-5 *Potentiometer controls length of pulse.*

SUMMARY

■ The output frequency of the timer is determined by the values of capacitor C_1 and resistors R_1 and R_2.

■ It will operate at almost any frequency desired.

■ If has found use in many industrial electronic circuits that require the use of a square-wave oscillator.

UNIT 34 REVIEW QUESTIONS

1. Why must pin #2 of the 555 timer be connected to pin #6? _____
_____

2. Assuming pin #5 is not being used, what controls the amount of time the timer is on? Refer to the figure below. _____
_____

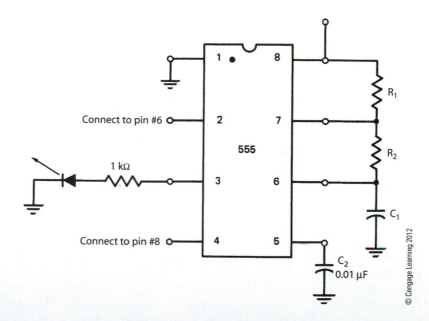

3. Assuming pin #5 is not being used, what controls the amount of time the timer remains off? Refer to the figure above. _____
_____

4. If pin #5 is connected to Vcc through a resistor, what effect does it have on the on time? _____
_____

5. If pin #5 is connected to ground through a resistor, what effect does it have on the on time? _____
_____
_____

6. What effect does pin #5 have on the off time of the timer? _____
_____

UNIT 35

THE 555 ON-DELAY TIMER

OBJECTIVES

After studying this unit, the student should be able to:

- Describe the operating characteristics of an on-delay timer.

- Discuss the operation of a stealer transistor.

- Construct a circuit using the 555 timer as an on-delay timer.

The 555 timer can be used to construct an on-delay relay. The 555 provides accurate time delays that can range from seconds to hours. These delays depend on the values of resistance and capacitance used in the circuit.

In the circuit shown in Figure 35-1, transistor Q_1 is used to switch relay coil K_1 on or off. The 555 timer may not be able to supply the current needed to operate the relay, so a transistor will be used to do the job of controlling the relay.

OPERATION OF THE TIMER

Transistor Q_2 is used as a stealer transistor to steal the base current from transistor Q_1. As long as transistor Q_2 is turned on by the output of the timer, transistor Q_1 is turned off.

Capacitor C_3 is connected from the base of transistor Q_1 to ground. It acts as a short time-delay circuit. When Vcc is first turned on by switch S_1, capacitor C_3 is in a discharged state. Before transistor Q_1 can turn on, capacitor C_3 must be charged through resistor R_3. This charging time is only a fraction of a second. It ensures that transistor Q_1 will not turn on before the output of the timer can turn on transistor Q_2. Once transistor Q_2 is turned on, it holds transistor Q_1 off by stealing its base current.

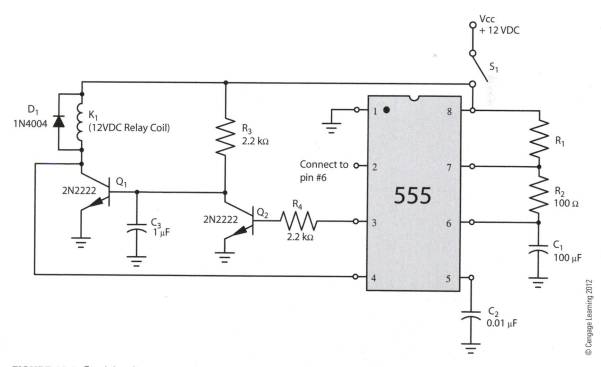

FIGURE 35-1 *On-delay timer*

Diode D_1 is used as a kick back or free-wheeling diode. It kills the spike voltage induced into the coil of relay K_1 when switch S_1 is opened. Resistor R_3 limits the base current to transistor Q_1. Resistor R_4 limits the base current to transistor Q_2.

Pin #4 is used as a latch in this circuit. When power is first applied at Vcc, transistor Q_1 is turned off. Because Q_1 is off, most of the applied voltage will be dropped across the transistor. This makes about 12 volts appear at the collector of the transistor. Pin #4 is connected to the collector so 12 volts is applied to pin #4. Pin #4 (the reset pin) must be connected to a voltage greater than two-thirds of Vcc if the timer is to operate. When it is connected to a voltage less than one-third of Vcc, it turns on the discharge and keeps the timer from operating. When transistor Q_1 turns on, the collector of the transistor drops to ground, or zero volt. Pin #4 is also connected to ground, preventing the timer from further operation. Because the timer cannot operate, the output remains off, and transistor Q_1 remains on.

Capacitor C_1 and resistors R_1 and R_2 are used to set the amount of time delay. (Note: Resistor R_2 should be kept at a value of about 100 Ω. Its job is to limit the current when capacitor C_1 discharges.) Resistor R_2 was made of a relatively low value to enable capacitor C_1 to discharge quickly. The value of resistor R_1 is changed to adjust the time setting.

Breaking the Operation Down

The circuit operates as follows:

1. Assume switch S_1 is open and all capacitors are discharged.

2. When switch S_1 closes, pin #2 is connected to zero volt and triggers the timer.

3. The output of the timer turns on transistor Q_2, which steals the base current of transistor Q_1. Q_1 remains off as long as Q_2 is on.

4. When capacitor C_1 is charged to two-thirds of Vcc, the discharge turns on and the output of the timer turns off.

5. When the output turns transistor Q_2 off, transistor Q_1 is supplied with base current through resistor R_3 and turns on relay coil K_1.

6. With transistor Q_1 on, the voltage applied to the reset pin, #4, changes from 12 volts to 0 volts.

7. When the reset is taken to 0 volts, the discharge is locked on and the output off.

8. Once transistor Q_1 has turned on, switch S_1 has to be reopened to reset the circuit.

UNIT 35 REVIEW QUESTIONS

Refer to Figure 35-1 to answer the following questions.

1. What function does transistor Q_1 serve in this circuit? _____

_____

_____

2. What function does diode D_1 serve in this circuit? _____

_____

_____

3. What function does transistor Q_2 serve in this circuit? _____

_____

_____

4. What is the function of pin #4 on the 555 timer in this circuit? _____

_____

5. What is the function of capacitor C_3 in this circuit? _____

_____

6. What sets the amount of time delay in this circuit? _____

_____

UNIT 36

THE 555 PULSE TIMER

OBJECTIVES

After studying this unit, the student should be able to:

- Describe the operation of a pulse timer.

- Discuss the operation of a blocking diode.

- Construct a pulse timer using a 555 timer.

In the circuit shown in Figure 36-1, the 555 is used as a pulse timer. It is connected differently than the other 555 timer circuits that have been covered. In this circuit, capacitor C_1 is charged by a feedback circuit from the output of the timer and not from Vcc.

- Transistor Q_1 is used to turn relay coil K_1 on or off.
- Diode D_1 is used as a kick back or free-wheeling diode.
- Resistor R_1 limits the current to the base of the transistor Q_1.
- Capacitor C_2 is used to disable pin #5 to ground.
- Diode D_2 is a blocking diode.
- Once capacitor C_1 has been charged, it is prevented from discharging through resistors R_2 and R_1; it must discharge through resistors R_3 and R_4.

In this circuit, pin #2 is connected to pin #6, which is at ground potential. This voltage is less than one-third of Vcc, so the output at pin #3 turns on when switch S_1 closes. With pin #3 on, two things happen: transistor Q_1 turns on and energizes relay coil K_1, and capacitor C_1 begins charging through resistor R_2. When capacitor C_1 is charged to two-thirds of Vcc, the output turns off, which turns off

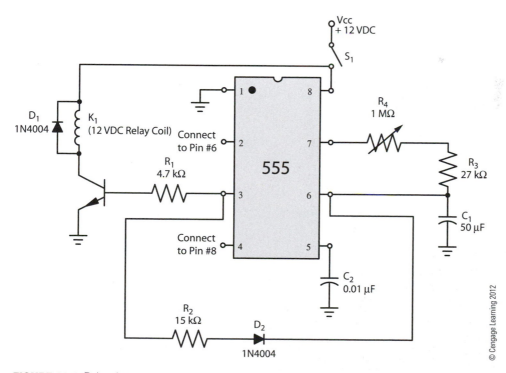

FIGURE 36-1 *Pulse timer*

relay K_1. The value of resistor R_2 determines how long relay K_1 stays on before it turns off.

When capacitor C_1 has been charged to two-thirds of Vcc, and the output has turned off, it discharges through resistors R_3 and R_4. When capacitor C_1 has been discharged to a point that its voltage is one-third of Vcc, pin #2 again triggers and turns the output on. The value of resistors $R_3 + R_4$ determines how long relay coil K_1 remains off before it turns on again. If the time between pulses is a fixed value, resistor R_4 is not necessary. If R_4 is a variable resistor, however, it will control the time between pulses.

UNIT 36 REVIEW QUESTIONS

Refer to Figure 36-1 to answer the following questions.

1. What is the function of diode D_2? _____

_____

2. Why must pin #4 be connected to pin #8? _____

_____

3. What determines the amount of time the timer will stay on before it turns off? _____

_____

4. What determines the amount of time the timer will stay off before it turns back on? _____

_____

5. What function does resistor R_1 serve in this circuit? _____

_____

UNIT 37

ABOVE- AND BELOWGROUND POWER SUPPLIES

OBJECTIVES

After studying this unit, the student should be able to:

- Describe an above- and belowground power supply.
- Discuss different methods of constructing an above- and belowground power supply.
- Discuss filtering for an above- and belowground power supply.
- Construct an above- and belowground power supply.

Most electronic circuits require a DC voltage to operate. A power supply that can furnish a positive and negative output voltage is sufficient for most circuits. However, there are some that require voltages that are above- and belowground.

DEFINING THE TERMS

In an electronic circuit, ground is considered to be 0 volts. Other values of voltage are referenced to ground. If a voltage is positive with respect to ground, it is considered to be *aboveground*. If a voltage is negative with respect to ground, it is considered to be *belowground*, Figure 37-1. Voltmeters V_1 and V_2 are both zero-center voltmeters. Both have their negative terminals connected to ground and their positive terminals connected to an input line.

Applying the Definitions

Notice that the pointer on voltmeter V_1 has moved to the right of the zero center. This indicates that the + (positive) terminal of the voltmeter is connected to a voltage that is positive with respect to ground. The pointer of voltmeter V_2, however, has moved to the left of the zero center, which indicates that its + terminal is connected to a voltage that is negative with respect to ground. If the + line is 12 volts above ground and the − (negative) line is 12 volts belowground, a voltmeter connected from the + line to the − line will indicate 24 volts.

CONSTRUCTION OF THE SUPPLY
Center-tapped

There are several ways to construct an above- and belowground power supply. One of the easiest is shown in Figure 37-2. The circuit is constructed by using a center-tapped transformer and a bridge

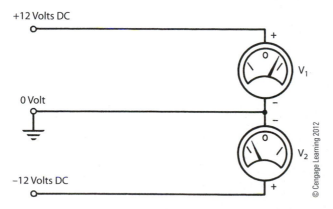

FIGURE 37-1 *Above- and belowground voltages*

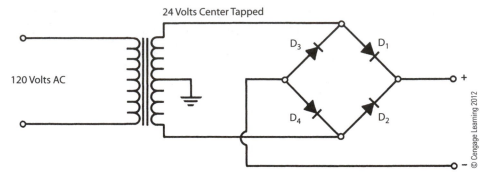

FIGURE 37-2 *An above- and belowground power supply*

FIGURE 37-3 *A two-diode full-wave rectifier producing a + (positive) voltage when compared to ground*

FIGURE 37-4 *A two-diode full-wave rectifier producing a − (negative) voltage when compared to ground*

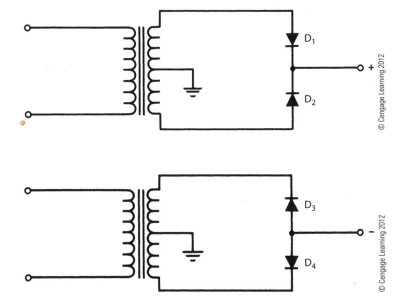

rectifier. Examine this circuit and notice that there are 2 two-diode-type full-wave rectifiers. Diodes D_1 and D_2 form one rectifier that produces a positive voltage with respect to ground, Figure 37-3. Diodes D_3 and D_4 form a two-diode type full-wave rectifier that produces a negative voltage when compared to ground, Figure 37-4.

The above- and belowground power supply shown in Figure 37-2 produces voltages that are equal when compared to ground. For instance, if the positive voltage is 12 volts aboveground, the negative voltage is 12 volts belowground. Another method for producing an above- and belowground power supply with equal voltages is shown in Figure 37-5.

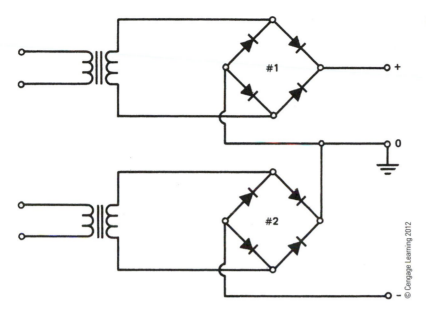

© Cengage Learning 2012

FIGURE 37-5 *A two-bridge rectifier circuit*

Two Transformers

This circuit uses two separate transformers and two bridge rectifiers. The negative output voltage of rectifier #1 is connected to the positive output voltage of rectifier #2. The point where this connection is made is the ground terminal. The positive output of rectifier #1 is the positive, or aboveground, voltage. The negative output of rectifier #2 is the negative, or belowground, voltage. This circuit can be used when a center-tapped transformer is not available, or when a higher voltage is needed than is available with one transformer.

The circuit in Figure 37-5 illustrates a method of obtaining an above- and belowground power supply using two transformers. The same results can be achieved by a simpler method, however.

Series Aided

Instead of using two bridge rectifiers, the secondary winding of the transformers can be connected series aiding, as shown in Figure 37-6. The connection point of the two transformer secondaries can be used as the center tap. Only one bridge rectifier is required in this case. This circuit operates the same as the circuit shown in Figure 37-2.

For small power applications, the circuit shown in Figure 37-7 can be used to obtain above- and belowground voltages. Diodes D_1 and D_2 are zener diodes of equal value. The applied DC voltage must

FIGURE 37-6 *Two transformers connected to form a center-tapped winding*

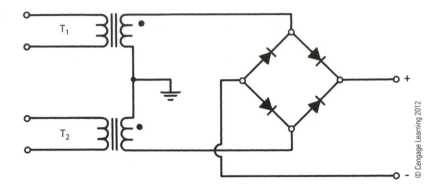

FIGURE 37-7 *Above- and belowground voltage produced by zener diode*

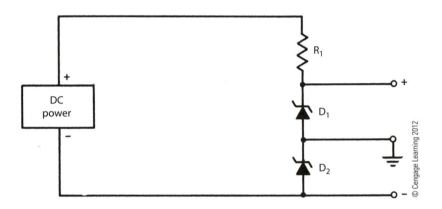

be greater than the combined voltage of the two zener diodes. For instance, if the zeners are rated at 12 volts each, the applied voltage must be greater than 24 volts. Resistor R_1 limits the current flow through the circuit. Its value must be set to permit enough current flow to operate the load. Because each zener diode has a voltage drop of 12 volts,

1. the connection point of the two diodes will be 12 volts more negative than the cathode of diode D_1,

2. and 12 volts more positive than the anode of diode D_2.

If the connection point of the two diodes is grounded, the cathode of diode D_1 is positive, or aboveground, and the anode of diode D_2 is negative, or belowground.

FILTERING

Above- and belowground power supplies are filtered the same way as any other power supply. A choke coil can be used to filter the current, and a capacitor to filter the voltage, Figure 37-8.

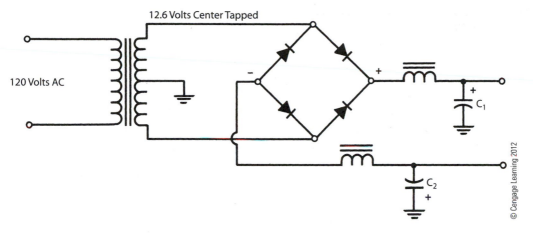

FIGURE 37-8 *Filtering an above- and belowground power supply*

Notice the connection of capacitor C_2 in Figure 37-8. The positive terminal has been connected to ground. Because the ground terminal of the power supply is more positive than the negative terminal, the polarity of the capacitor must be observed when this connection is made.

UNIT 37 REVIEW QUESTIONS

1. What is an above- and belowground power supply? _____
_____

2. What type of transformer must be used when using a bridge rectifier as an above- and belowground power supply? _____
_____

3. When using an above- and belowground power supply, is the positive voltage considered to be aboveground or belowground? _____
_____

4. Describe two methods of constructing an above- and belowground power supply using two transformers that do not have a center tap. _____
_____
_____
_____
_____
_____
_____

5. When connecting electrolytic capacitors as filters for above- and belowground power supplies, which terminal of the capacitor used to filter the negative voltage must be connected to ground?
_____
_____

UNIT 38

THE OPERATIONAL AMPLIFIER

OBJECTIVES

After studying this unit, the student should be able to:

- Discuss the operation of an operational amplifier.

- Discuss inverting inputs and noninverting inputs.

- Describe specific parameters for the 741 operational amplifier.

- Discuss the operation of the offset null.

- Discuss negative feedback and calculate the gain of the operational amplifier.

- Connect an operational amplifier in a circuit.

The operational amplifier, like the 555 timer, has become another very common component found in industrial electronic circuits. The operational amplifier (op amp) is used in hundreds of different applications. Op amps differ, depending on the circuit that they are intended to operate. Some use bipolar transistors and others use field effect transistors (FETs) for the input. FETs have extremely high input impedance. This can be several thousand megohms. The advantage of this impedance is that a large amount of current is not needed to operate the amplifier. Op amps that use FET inputs are generally considered as requiring no input current.

THE IDEAL OP AMP

There are three things that would go into the making of an ideal amplifier:

1. The ideal amplifier should have an input impedance of infinity. If the amplifier has such an impedance, it requires no power drain on the signal source that is to be amplified. Therefore, regardless of how weak the input signal source is, it will not be affected when connected to the amplifier.

2. The ideal amplifier should have zero output impedance. If it has zero output impedance, it can be connected to any load resistance desired. No voltage will drop inside the amplifier. If there is no internal voltage drop, the amplifier will utilize 100 percent of its gain.

3. The amplifier will have unlimited gain. This permits it to amplify any input signal as much as desired.

THE 741 OP AMP

There is no such thing as the ideal or perfect amplifier, of course, but the op amp comes close. One of the op amps that is still used in industry is the 741. It will be used in this unit as a typical operational amplifier. There are other op amps that have different characteristics of input and output impedance, but the basic theory of operation is the same for all.

The 741 op amp uses bipolar transistors for the input. The input impedance is about 2 MΩ, and the output impedance is about 75 Ω. Its open loop (maximum gain) is about 200,000. The 741 op amp has such a high gain that it is generally impractical to use. Negative feedback, which will be discussed later in the unit, is used to reduce the gain. Assume the amplifier has an output voltage of 15 volts. If the input signal voltage is greater than 1/200,000 of the output voltage or 75 microvolts (15/200,000 = 0.000075), the amplifier will be driven into saturation.

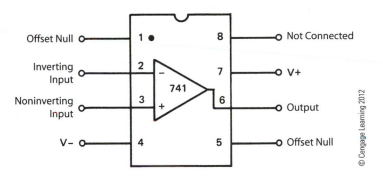

FIGURE 38-1 *The 741 operational amplifier*

Operating the 741

The 741 operational amplifier is generally housed in an 8-pin inline IC package, Figure 38-1. Pins #1 and #5 are connected to the offset null that can be used to produce 0 volts at the output. This is the process:

1. The op amp has two inputs: the inverting and the noninverting input.

2. They are connected to a differential amplifier that amplifies the difference between the two voltages.

3. If both of these inputs are connected to the same voltage (by grounding both inputs), the output should be 0 volts.

In actual practice, however, there are generally unbalanced conditions in the op amp that cause a voltage to be produced at the output. Because the op amp has a very high gain, any slight imbalance of a few microvolts at the input can cause several millivolts at the output. The offset nulls are adjusted after the 741 is connected into a working circuit. Adjustment is made by connecting a 10 kilohm (kΩ) potentiometer across pins #1 and #5, and connecting the wiper to the negative voltage, Figure 38-2.

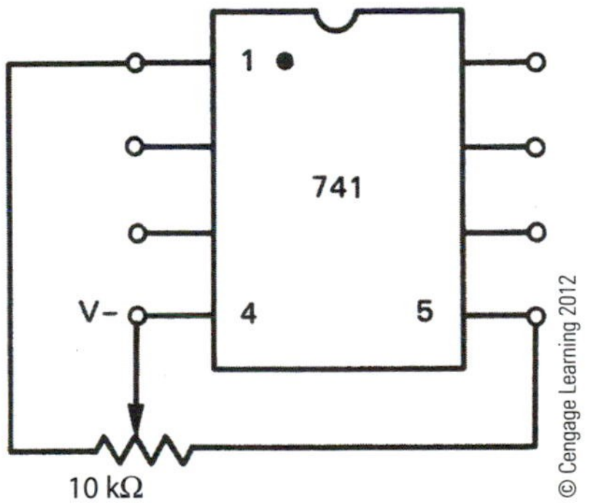

FIGURE 38-2 *Offset null connection*

Pin Explanation

Pin #2 is the inverting input. If a signal is applied to this input, the output will be inverted. A positive-going AC voltage applied to the inverting input will produce a negative-going output voltage, Figure 38-3.

FIGURE 38-3 *Inverted output*

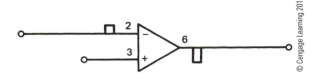

FIGURE 38-4 *Noninverted output*

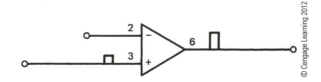

© Cengage Learning 2012

Pin #3 is the noninverting input. When a signal voltage is applied to the noninverting input, the output voltage will be the same polarity. A positive-going AC signal applied to the noninverting input will produce a positive output voltage, Figure 38-4.

Pins #4 and #7 are the voltage input pins. Operational amplifiers are generally connected to above- and belowground power supplies. There are some circuit connections that do not require an above- and belowground power supply, but these are the exception instead of the rule. Pin #4 is connected to the negative (belowground) voltage and pin #7 is connected to the positive (aboveground) voltage. The 741 will operate on voltages that range from about 4 to 16 volts. The operating voltage is usually 12–15 volts plus and minus. The 741 has a maximum power output rating of about 500 mW. Pin #6 is the output pin, and #8 is not connected.

Open Loop Gain

As stated before, the open loop gain of the 741 operational amplifier is about 200,000. This is not practical for most applications, so this gain must be reduced to a reasonable level. One of the op amp's greatest advantages is the ease with which the gain can be controlled, Figure 38-5. The amount of gain is controlled by a negative feedback loop; a portion of the output voltage is fed back to the inverting input. This is so because

1. the output voltage is always opposite in polarity to the inverting input voltage,

2. the amount of output voltage fed back to the input tends to reduce the input voltage.

Negative feedback has two effects on the operation of the amplifier. It reduces the gain, and it makes the amplifier more stable.

FIGURE 38-5 *Negative feedback connection*

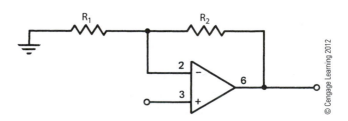

© Cengage Learning 2012

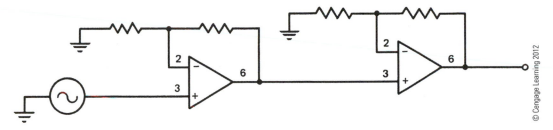

FIGURE 38-6 *Using more than one op amp to obtain higher gain*

The gain of the amplifier is controlled by the ratio of resistors R_2 and R_1. If a noninverting amplifier is used, the gain is found by the formula

$$\frac{(R_2 + R_1)}{R_1}$$

If resistor R_1 is 1 kΩ and resistor R_2 is 10 kΩ, the gain of the amplifier is 11 (11,000/1000 = 11).

If the op amp is connected as an inverting amplifier, however, the input signal will be out-of-phase with the feedback voltage of the output. This causes a reduction of the input voltage applied and a reduction in gain. The formula R_2/R_1 is used to compute the gain of an inverting amplifier. If resistor R_1 is 1 kΩ and resistor R_2 is 10 kΩ, the gain of the inverting amplifier is 10 (10,000/1000 = 10).

There are some practical limits, however. As a rule, the 741 operational amplifier is not operated above a gain of 100. If more gain is desired, it is generally obtained by using more than one amplifier, Figure 38-6.

As shown in the figure, the output of one amplifier is fed into the input of another. The 741 is not operated at high gain because at high gains it tends to become unstable. Another rule for operating the 741 op amp is that the total feedback resistance, $R_1 + R_2$, is usually kept at more than 1000 Ω and less than 100,000 Ω. These rules apply to the 741 operational amplifier but may not apply to other op amps.

USING THE OP AMP

Op amps are generally used in three basic ways. This does not mean that they are used in only three circuits, but that there are three basic circuits used to build other circuits.

One of these is the voltage follower. In this circuit, the output of the op amp is connected directly back to the inverting input, Figure 38-7. Because there is a direct connection between the output

FIGURE 38-7 *Voltage follower connection*

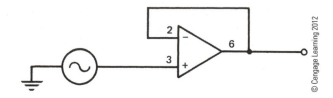

FIGURE 38-8 *Noninverting amplifier connection*

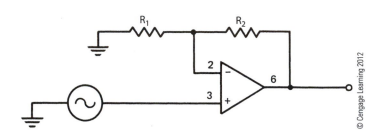

of the amplifier and inverting input, the gain of this circuit is one. For instance, if a signal voltage of 0.5 volt is connected to the noninverting input, the output voltage will also be 0.5 volt. This circuit amplifies the input impedance by the amount of the open loop gain. If the 741 has:

■ an open loop gain of 200,000 and

■ an input impedance of 2 MΩ, the circuit will give the amplifier an input impedance of 200 kΩ × 2 MΩ, or 400,000 MΩ.

The circuit connection is generally used for impedance matching purposes.

The second basic circuit is the noninverting amplifier, Figure 38-8. In this circuit, the output voltage is the same polarity as the input voltage. If the input voltage is a positive-going voltage, the output is a positive-going voltage. The amount of gain is set by the ratio of resistors $R_1 + R_2/R_1$ in the negative feedback loop.

The third basic circuit is the inverting amplifier, Figure 38-9. In this circuit, the output voltage is opposite in polarity to the input

FIGURE 38-9 *Inverting amplifier connection*

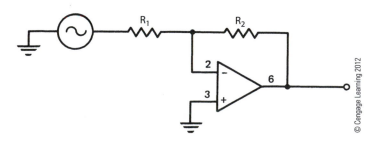

voltage. If the input signal is a positive-going voltage, the output voltage is negative going at the same instant in time. The gain of the circuit is determined by the ratio of resistors R_2/R_1.

APPLICATIONS

In the circuit shown in Figure 38-10, an operational amplifier is used as a timer. Resistors R_1 and R_2 form a voltage divider that supplies one-half the input voltage to the inverting input. Resistor R_3 and capacitor C_1 form an RC time constant. When switch S_1 is closed, the noninverting input is held at ground potential, which forces the output to remain low. When switch S_1 is opened, capacitor C_1 begins to charge through resistor R_3. When the voltage applied to the noninverting input reaches a value greater than that applied to the inverting input, the output goes high. The amount of time delay is determined by the values of C_1 and R_3. By choosing the correct values, the timer can have a delay that can range from seconds to hours. The time delay is approximately 0.7 CR second where the value of C is in farads.

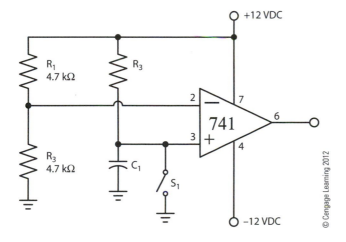

FIGURE 38-10 *Operational amplifier used as a timer*

© Cengage Learning 2012

UNIT 38 REVIEW QUESTIONS

1. What is the advantage of using field effect transistors as inputs for an operational amplifier?

_____

_____

2. What is the input impedance of a 741 op amp? _____

_____

3. How is the gain of an op amp controlled? _____

_____

4. What two effects do negative feedback have on an operational amplifier circuit? _____

_____

5. Resistor R_1 of a noninverting amplifier has a resistance of 750 Ω and resistor R_2 has a resistance of 15,000 Ω. What is the gain of this amplifier? _____

_____

6. Resistor R_1 of an inverting amplifier has a resistance of 1200 Ω and resistor R_2 has a resistance of 100,000 Ω. What is the gain of this amplifier? _____

_____

UNIT 39

THE 741 OP AMP LEVEL DETECTOR

OBJECTIVES

After studying this unit, the student should be able to:

- Describe the operation of a level detector.

- Connect an operational amplifier as a noninverting level detector.

- Connect an operational amplifier as an inverting level detector.

The operational amplifier is often used as a level detector or comparator. In the circuit shown in Figure 39-1, the 741 op amp will be used as an inverted amplifier to detect when one voltage becomes greater than another.

USING THE OP AMP AS AN INVERTED AMPLIFIER

Notice that this circuit does not use an above- or belowground power supply. Instead, it is connected to a power supply with a single positive and negative output. During normal operation, the noninverting input of the amplifier is connected to a zener diode. The zener diode produces a constant positive voltage at the noninverting input of the amplifier and is used as a reference. As long as the noninverting input is more positive than the inverting input, the output of the amplifier will be high. An LED, D_1, is used to detect a change in the polarity of the output. As long as the output of the op amp remains high, the LED stays turned off. This is because the LED has equal voltage applied to both its anode and cathode. Because both are connected to +12 volts, there is no potential difference. Therefore, no current flows through the LED.

If the voltage at the inverting input becomes more positive than the reference voltage applied to pin #3, the output voltage will go low, to about +2.5 volts. The output voltage of the op amp will not go to zero, or ground, because it is not connected to a voltage that is below ground. If the output voltage is to go to 0 volts, pin #4 must be connected to a voltage that is below ground. When the output goes low, there is a potential of about 9.5 volts (12 − 2.5 = 9.5) produced across R_1 and D_1. This causes the LED to turn on and indicate that the state of the op amp's output has changed from high to low.

AS A DIGITAL DEVICE?

In this circuit, the op amp appears to be a digital device because the output seems to have only two states, high or low. Actually, it is not

FIGURE 39-1 *Inverting level detector*

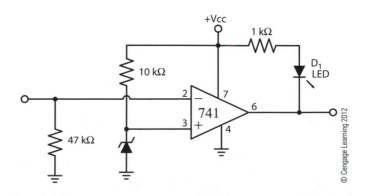

a digital device: this circuit only makes it appear to be digital. Notice in Figure 39-1 that there is no negative feedback loop connected between the output and the inverting input. The amplifier uses its open loop gain (which is about 200,000 for the 741) to amplify the voltage difference between the inverting and the noninverting input. If

1. the voltage applied to the inverting input becomes 1 millivolt more positive than the reference voltage applied to the noninverting input,

2. then the amplifier will try to produce an output that is 200 volts more negative than its high-state voltage $(0.001 \times 200,000 = 200)$.

The output voltage of the amplifier cannot be driven 200 volts more negative, of course. Because there are only 12 volts applied to the circuit, the output voltage simply reaches the lowest voltage it can and then goes into saturation. As shown, the op amp is not a digital device, but it can be made to act like one.

Replacing the Zener with a Voltage Divider

If the zener diode is replaced with a voltage divider, as shown in Figure 39-2, the reference voltage can be set to any value desired. By adjusting the variable resistor in the figure, the positive voltage applied to the noninverting input can be set the same way. If the voltage at the noninverting input is set for 3 volts, the output of the op amp will go low when the voltage applied to the inverting input becomes greater than +3 volts. If the voltage connected to the non-inverting input is changed to +8 volts, the output voltage will go low when the inverting input voltage becomes greater than +8 volts. This circuit permits the voltage level at which the output of the op amp will change to be adjusted.

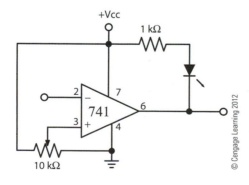

FIGURE 39-2 *Adjustable inverting level detector*

CHANGING THE OUTPUT FROM LOW TO HIGH

In these two circuits, the op amp changed from a high level to a low level when activated. There may be occasions, however, when it is desirable to change the output from a low to a high level. This can be done by connecting the inverting input to the reference voltage and the noninverting input to the voltage being sensed, Figure 39-3.

In this circuit, the zener diode provides a positive reference voltage to the inverting input. As long as this voltage remains more positive than the voltage at the noninverting input, the output voltage of the op amp will remain low. If the voltage applied to the noninverting input becomes more positive than the reference voltage, the output of the op amp will become high.

A Minor Problem with the Circuit

Depending on the application, this circuit could cause a minor problem. Because this circuit does not use an above- and belowground power supply, the low output voltage of the op amp will be about +2.5 volts. This positive voltage could cause any devices connected to the output to turn on even if they should be off. If the LED in Figure 39-3 is used, it will glow dimly when the output is in the low state. This problem can be corrected to two ways.

SOLUTION: One way is to connect the op amp to an above- and belowground power supply, as shown in Figure 39-4. The output voltage of the op amp in this circuit is negative (belowground) as long as the voltage applied to the inverting input is more positive than the voltage applied to the noninverting input. As long as the output voltage of the op amp is negative with respect to ground, the LED is reverse biased and cannot operate. When the voltage applied to the noninverting input becomes more positive than the voltage applied to the inverting input, the output of the op amp becomes positive and the LED turns on.

FIGURE 39-3 *Noninverting level detector*

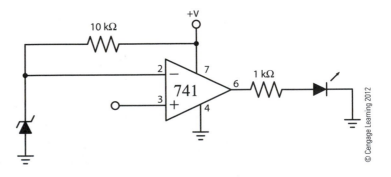

© Cengage Learning 2012

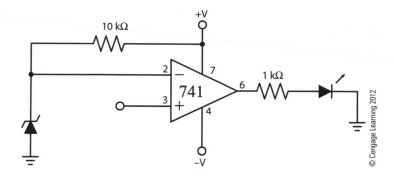

FIGURE 39-4 *Belowground power connection permits the output voltage to become negative.*

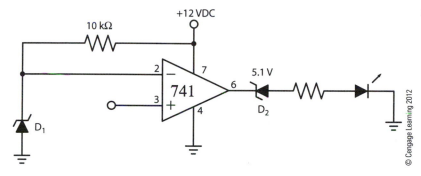

FIGURE 39-5 *A zener diode is used to keep the output turned off.*

ANOTHER SOLUTION: The second method of correcting the output voltage problem is shown in Figure 39-5. In this circuit, the op amp is connected to a power supply that has a single positive and negative output. A zener diode, D_2, has been connected in series with the output of the op amp and the LED. The voltage value of diode D_2 is greater than the output voltage of the op amp in the low state, but less than the output voltage of the op amp in its high state. Assume that the value of zener diode D_2 is 5.1 volts. If the output voltage of the op amp in its low state is 2.5 volts, D_2 is turned off and will not conduct. If the output voltage becomes +12 volts (in the op amp's high state), the zener diode turns on and conducts current to the LED. The zener diode, D_2, keeps the LED turned completely off until the op amp switches to its high state. It then provides enough voltage to overcome the reverse voltage drop of the zener diode.

SUMMARY

In the preceding circuits, an LED was used to indicate the output state of the amplifier. Keep in mind that the LED is used only as a detector and that the output of the op amp can be used to control almost anything. For example, the output of the op amp can be connected to the base of a transistor, as shown in Figure 39-6. The

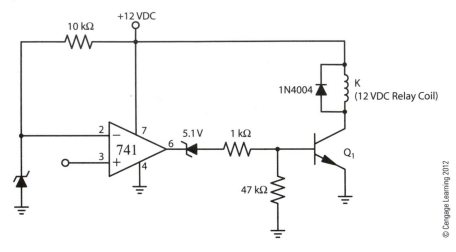

FIGURE 39-6 *A transistor is used to control a relay.*

transistor can then control the coil of a relay. The relay can be used to control almost anything.

TEMPERATURE DEPENDENT RESISTORS

In this circuit, the temperature of the water in the tank and the temperature of the collector are sensed by temperature-dependent resistors. Temperature-dependent resistors are devices that exhibit a change of resistance with a change of temperature. There are two basic types of temperature-dependent resistors. One type is called a **resistive temperature detector**, or RTD. RTDs are constructed of metal and have a *positive temperature coefficient.* The temperature coefficient is the ratio of change in resistance as compared to a change in temperature. The word *positive* means that the resistance of the device will increase as the temperature increases. The temperature coefficient of an RTD is determined by the metal of which it is constructed. As a general rule, RTDs will have a linear coefficient of temperature. The characteristic curve of an RTD will approximate a straight line, as shown in Figure 39-7. The disadvantage of the RTD is that it has a low temperature coefficient. It does not exhibit a large change in resistance over a large change of temperature.

The second type of temperature dependent resistor is known as a **thermistor**. Thermistors are constructed of metallic oxides and exhibit a negative temperature coefficient. This means that their resistance will decrease with an increase of temperature. Thermistors have an advantage in that they have a much higher temperature coefficient than do RTDs, but they have a disadvantage in that they are

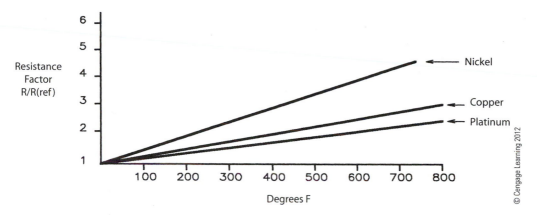

FIGURE 39-7 *Characteristic curves of different RTDs*

nonlinear. Figure 39-8 illustrates the characteristic curves of several different thermistors. Notice the resistance scale is logarithmically shown to permit higher resistance values to be indicated.

The schematic symbol for a temperature-dependent resistor is generally a resistor symbol drawn inside a circle with the letter T written beside it. An arrow through the resistor indicates the resistor is variable, Figure 39-9. Some schematic diagrams point the arrow toward the top of the circle to indicate the use of an RTD, and toward the bottom of the circle to indicate the use of a thermistor. This, however, is not a universally accepted practice.

In the circuit shown in Figure 39-10, the operational amplifier is connected to +12 volts at pin #7, and pin #4 is at ground potential. The output of the op amp is intended to operate the input of a solid-state relay, which in turn controls the operation of the pump motor. Because pin #4 has been connected to ground instead of a negative voltage, a 5.1-volt zener diode has been connected between the output of the op amp and the input of the solid-state relay. The zener diode prevents the relay from being turned on when the op amp is in the low state.

Thermistor T_2 and resistor R_3 form a voltage divider circuit for the noninverting input. Thermistor T_2 is placed in a position that will permit it to sense the temperature of the water in the storage tank. Thermistor T_1 and resistors R_1 and R_2 form a voltage divider circuit for the inverting input. Thermistor T_1 is attached to the collector, which permits it to sense the collector's surface temperature. Resistor R_2 is variable to permit a range of adjustment. It is the setting of this resistor that will determine the difference in temperature that is necessary between the collector and the tank to turn on the pump.

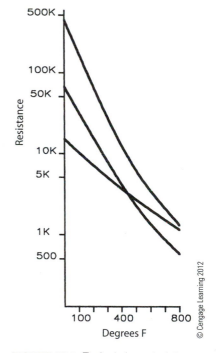

FIGURE 39-8 *Typical characteristic curves of thermistors*

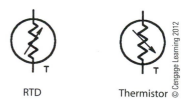

RTD   Thermistor

FIGURE 39-9 *Schematic symbols for temperature dependent resistors*

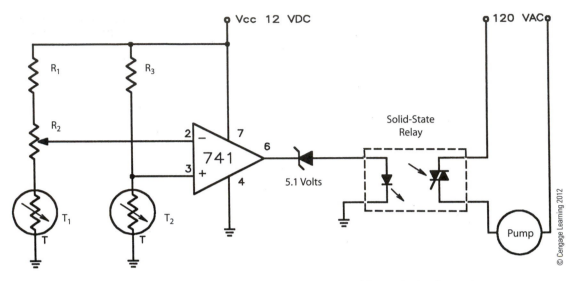

FIGURE 39-10 *Circulating pump control for a solar heating system*

To understand the operation of the circuit, assume that the water in the tank is at a higher temperature than the surface of the collector. Resistor R_2 has been adjusted to permit the voltage applied to the inverting input to be higher than the voltage applied to the noninverting input. The output of the op amp is low and the pump motor is turned off. Now assume that the temperature of the collector begins to increase. This increase of temperature causes the resistance of thermistor T_1 to decrease. As the resistance of T_1 decreases, the voltage applied to the inverting input decreases also. When the voltage applied to the inverting input becomes lower than the voltage applied to the noninverting input, the output of the op amp turns on and starts the pump motor.

As water is calculated from the storage tank to the collector, it will increase in temperature. The increased water temperature will cause the resistance of T_2 to decrease and lower the voltage applied to the noninverting input. When the water has been heated to within a few degrees of the collector temperature, the resistance of T_2 becomes low enough to permit the output of the op amp to turn off.

APPLICATIONS

A good example of how an operational amplifier can be used as a level detector can be seen in Figure 39-10. This circuit is used to control the operation of a pump motor that circulates water from a storage tank to a solar collector, as shown in Figure 39-11. Pump operation

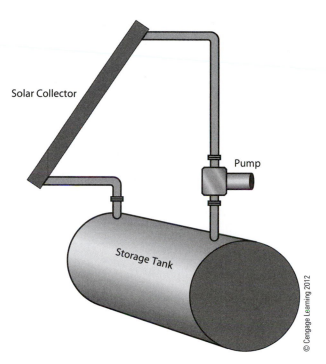

FIGURE 39-11 *Pump circulates water from tank to collector.*

Solar Collector

Pump

Storage Tank

© Cengage Learning 2012

is determined by the difference in temperature between the water in the tank and the collector. If the water in the tank is at a higher temperature than the surface of the collector, the pump motor is turned off. When the collector temperature becomes greater than the tank temperature, the pump turns on and circulates water from the tank to the collector.

UNIT 39 REVIEW QUESTIONS

1. A 741 operational amplifier is to be used as a level detector. A zener diode with a voltage rating of 4.7 volts is connected to the noninverting input. When the voltage applied to the inverting input becomes more than 4.7 volts positive, will the output become high or low? _____

_____

2. If the 741 is not connected to an above- and belowground power supply but is connected to a 12-volt DC power supply with a positive and negative output, what will be the output voltage of the op amp in the low state? _____

_____

3. What is the advantage of using a voltage divider circuit to supply the reference voltage instead of a zener diode? _____

_____

4. A zener diode with a voltage rating of 6 volts is connected to the inverting input. When the voltage applied to the noninverting input becomes more than +6 volts, will the output become high or low?

_____

_____

5. An op amp being used as a level detector has an output voltage of 3 volts in the low state and 15 volts in the high state. Can a zener diode with a voltage rating of 18 volts be used on the output of the op amp to keep it turned off during the time the op amp is in the low state? _____

_____

THE 741 OPERATIONAL AMPLIFIER USED AS AN OSCILLATOR

OBJECTIVES

After studying this unit, the student should be able to:

- Discuss the operation of an oscillator.

- Describe the difference between an oscillator and a pulse generator.

- Construct an oscillator using a 741 operational amplifier.

- Construct a pulse generator using a 741 operational amplifier.

An operational amplifier can be used as an oscillator. The circuit shown in Figure 40-1 is a very simple circuit that produces a square-wave output.

THE CIRCUIT

This circuit is rather impractical, however. The circuit depends on a slight imbalance in the op amp or random circuit noise to start the oscillator. Recall from previous units that a voltage difference of only a few millivolts is all that is needed to make the output of the amplifier go high or low. For example, if the inverting input becomes slightly more positive than the noninverting input, the output goes low (negative). With the output negative, capacitor C_t charges through resistor R_t to the negative value of the output voltage. When the voltage applied to the inverting input becomes slightly more negative than the noninverting input, the output changes to a high (positive) value of voltage. With the output positive, capacitor C_t charges through resistor R_t toward the positive output voltage. This circuit works quite well if

1. the op amp has no imbalance, and

2. the op amp is shielded from all electrical noise.

In practical application, however, there is generally enough imbalance in the amplifier or enough electrical noise to send the op amp into saturation. This stops the operation of the circuit.

Adding a Hysteresis Loop

The real problem with this circuit is that a millivolt difference between the two inputs is enough to drive the amplifier's output from one state to the other. This can be corrected with the addition of a **hysteresis loop** connected to the noninverting input as shown in Figure 40-2.

FIGURE 40-1 *Simple square-wave oscillator*

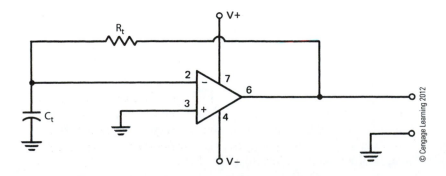

© Cengage Learning 2012

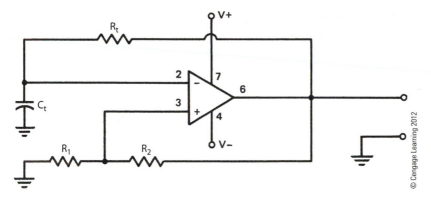

FIGURE 40-2 *Square-wave oscillator using a hysteresis loop*

OPERATION OF THE CIRCUIT

Resistors R_1 and R_2 form a voltage divider for the noninverting input. They are generally of equal value. To understand the circuit operation:

1. Assume that the inverting input is slightly more positive than the noninverting input.

  ■ This causes the output voltage to go negative.

2. Also assume that the output voltage is now negative 12 volts as compared to ground.

  ■ If the resistors are of equal value, the noninverting input is driven to −6 volts by the voltage divider.

Capacitor C_t begins to charge through resistor R_t to the value of the negative output voltage. When C_t has charged to a value slightly more negative than the −6 volts at the noninverting input, the op amp's output goes high (to +12 volts aboveground). At this point, the voltage applied to the noninverting input changes from −6 to +6 volts. Capacitor C_t begins to discharge through R_t to the positive voltage of the output. When the voltage applied to the inverting input becomes more positive than the noninverting input, the output changes to a low value (−12 volts). The voltage at the noninverting input is driven from +6 to −6 volts, and capacitor C_t again begins to charge toward the negative output voltage.

The addition of the hysteresis loop greatly changes the operation of the circuit. The differential between the two inputs is now volts instead of millivolts. The output frequency of the oscillator is determined by the values of C_t and R_t. The period of one cycle can be computed by using the formula: $T = 2RC$.

THE OP AMP AS A PULSE GENERATOR

The operational amplifier can also be used as a pulse generator. The difference between an oscillator and a pulse generator is the period of time that the output remains on as compared to the time it remains low or off.

An oscillator produces a waveform that has positive and negative pulses of equal voltage and time. In Figure 40-3, notice that the positive and negative values of voltage are the same. Both the positive and negative cycles remain turned on for the same amount of time. This waveform is the same as the one seen when an oscilloscope is connected to the output of a square-wave oscillator.

If the oscilloscope is connected to a pulse generator, however, a waveform similar to the one shown in Figure 40-4 will be seen. The positive value of the voltage is the same as the negative, just as in Figure 40-3. However, the positive pulse is much shorter than the negative pulse.

The 741 operational amplifier can easily be changed from a square-wave oscillator to a pulse generator. The circuit shown in Figure 40-5 is the same as the square-wave oscillator with the addition of resistor R_3 and R_4, and diodes D_1 and D_2.

FIGURE 40-3 *Output of an oscillator*

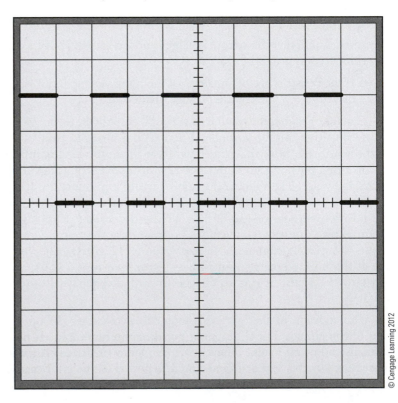

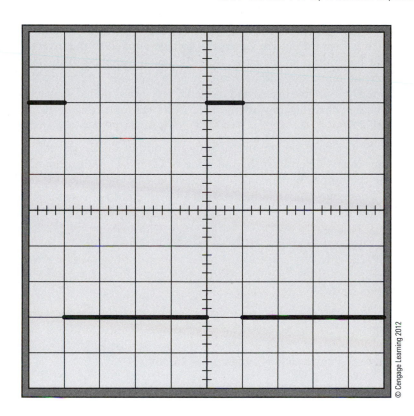

FIGURE 40-4 *Output of a pulse generator*

© Cengage Learning 2012

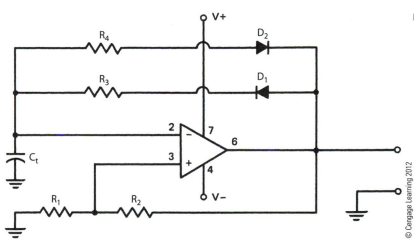

FIGURE 40-5 *Pulse generator circuit*

© Cengage Learning 2012

It permits capacitor C_t to charge at a different rate when the output is high (positive) than when the output is low (negative). Assume that the voltage of the op amp's output is low (-12 volts). Because the output voltage is negative, D_1 is reverse biased and no current flows through R_3. Therefore, C_t must charge through R_4 and D_2, which is forward biased. When the voltage applied to the inverting input becomes more negative than the voltage applied to the non-inverting input, the output of the op amp becomes +12 volts. Now diode D_2 is reverse biased and diode D_1 is forward biased. Capacitor C_t begins charging toward the +12 volts through resistor R_2 and diode D_1.

Notice that the amount of time the output of the op amp remains low is determined by the value of C_t and R_4. The amount of time the output remains high is determined by the value of C_t and R_3. These two ratios can be determined by the ratio of resistor R_3 to resistor R_4.

UNIT 40 REVIEW QUESTIONS

1. Is it necessary to use an above- and belowground power supply when a 741 operational amplifier is to be used as an oscillator? _____

_____

2. Is the hysteresis loop connected to the inverting or noninverting input? _____

_____

3. The hysteresis loop is actually a(n) _____

_____

4. An op amp has a capacitor with a value of 0.1 μF and a resistor with a value of 4.7 Ω connected to the inverting input. What would be the output frequency of the oscillator? _____

_____

_____

5. What is the difference between an oscillator and a pulse generator? _____

_____

_____

UNIT 41

VOLTAGE REGULATORS

OBJECTIVES

After studying this unit, the student should be able to:

- Describe the operation of a series regulator.

- Describe the operation of a shunt regulator.

- Discuss the need for voltage regulation in a circuit.

- Discuss current limit control.

- Construct a voltage regulated power supply with current limit.

A device used to change AC into DC voltage is generally referred to as a power supply. They range in complexity from a simple half-wave rectifier, as shown in Figure 41-1, to a unit that is voltage regulated, current limited, and temperature protected.

THE BATTERY AS A POWER SUPPLY

Many of the circuits used in industry are sensitive to a change in voltage. A regulator must be used to provide a constant output voltage. The voltage of an unregulated power supply changes with a change in load current. This change is caused by the internal impedance of the circuit. A nearly perfect DC power supply is probably the battery, but it, like all components, has some internal impedance. It is generally very low, but nevertheless, it's there.

Assume a 12-volt battery has an internal impedance of 0.1 Ω, Figure 41-2. If a 1.2-Ω resistor is connected to the battery terminals, 10 A of current will attempt to flow through the load. This current causes a voltage drop across the terminals of 1 volt ($10 \times 0.1 = 1$). Note that even a battery has some voltage drop when a load is connected to it. If the battery did not have internal impedance, it could produce unlimited current ($12/0 = $ infinity). Electronic power supplies are similar because they have internal impedance too.

Adding a Variable Resistor

This problem can be corrected, however. Assume that a DC power supply is needed that can furnish 2 A of current at 12 volts. To do this a power supply must first be constructed that can furnish more than 12 volts at 2 A, for example, 14 volts at 2.5 A. A variable resistor connected in series with the load can adjust the voltage at the output if the load current should change, Figure 41-3.

If the load connected to the output is small (0.1 A), resistance can be added in series to produce a voltage of 12 volts at the output. If the current is increased to 1.5 A, the resistance can be decreased to produce 12 volts. Note that the output voltage of the power supply can be adjusted (regulated) by the amount of series

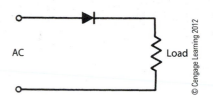

FIGURE 41-1 *Half-wave rectifier*

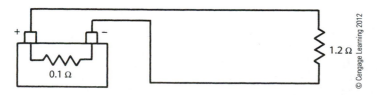

FIGURE 41-2 *Internal impedance*

FIGURE 41-3 *Series resistor used to regulate voltage to the load*

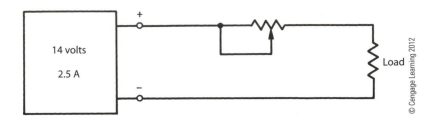

resistance in the circuit. This type of regulator isn't too practical, however, because

1. it requires the constant attention of an operator, and

2. people don't have reflexes that are fast enough to catch a sudden increase or decrease in load current.

THE ZENER DIODE AS A REGULATOR

The simplest electronic voltage regulator is the zener diode, which was covered in Unit 9. Almost all solid-state devices have a characteristic known as dynamic (changing) impedance.

A common junction diode has a forward voltage drop of 0.6 to 0.7 volts regardless of the amount of current flowing through it. In order for the drop to remain constant, the impedance of the device must change when current changes. Assume that a diode has a forward current of 0.05 A. Its impedance is 12 Ω (0.6/0.05 = 12). If the current is increased to 1 A, the impedance now appears to be 0.6 Ω (0.6/1 = 0.6).

The zener diode acts the same when connected in the reverse direction, Figure 41-4. Regardless of the current flowing through the zener in the reverse direction, its voltage remains constant.

The zener diode is generally used as a shunt regulator; it is connected in parallel with the load. The circuit shown in Figure 41-5 is generally used for low-power applications. Power zeners are available if more power is needed. A power transistor controlled by a zener diode can be used to handle the power of the circuit, Figure 41-6.

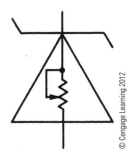

FIGURE 41-4 *Dynamic impedance*

Disadvantages of the Zener Diode

Shunt regulators are used in some applications, but they have some disadvantages. They require the use of a series resistor, Rs. This limits the current flow through the transistor when no load is connected to the output. It also adds internal impedance to the power supply, which puts a definite limit on the output current.

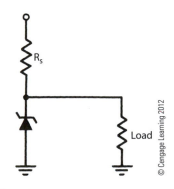

FIGURE 41-5 *Shunt regulator*

FIGURE 41-6 *Power zener circuit*

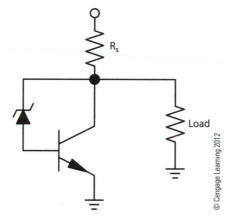

SERIES CONTROL AS A SOLUTION: Most voltage regulators use a series control, as shown in Figure 41-7. This type of regulator controls the output voltage by either

1. turning the transistor on harder if the voltage tries to drop, or

2. turning it off if the voltage tries to increase.

The zener diode connected between base and ground tries to keep the emitter voltage at the same potential as the base. If a load is connected between the emitter and ground, the voltage at the emitter will try to drop. When this happens, the zener diode begins to raise its internal impedance. This causes more base current to flow to the transistor. The increase of base current causes the transistor to turn on harder and raises the voltage of the emitter back to the voltage of the zener diode.

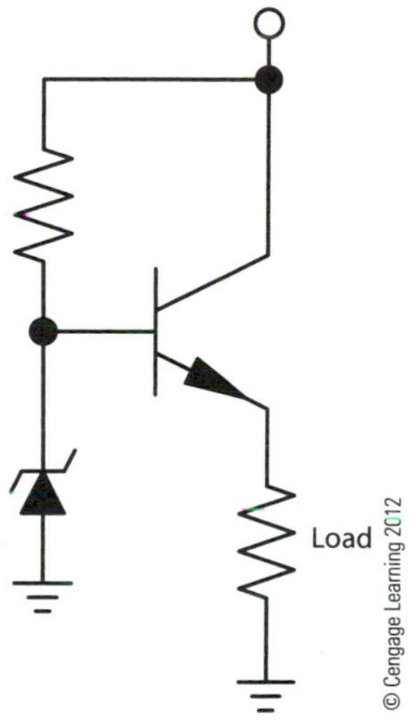

Although the circuit shown in Figure 41-7 works well on paper, it has a problem: the percentage of regulation is proportional to the amount of gain in the circuit, and power transistors are not famous for their gain.

THE OP AMP AS A VOLTAGE REGULATOR

Operational amplifiers, however, are famous for their gain. The op amp can be used to make a voltage regulator with a high percentage of regulation.

The operational amplifier shown in Figure 41-8 uses a zener diode to provide a reference voltage to the noninverting input. The inverting input is connected to the output of the op amp. Connecting the inverting input directly to the output forces the output to assume the same voltage as that applied to the noninverting input. Remember: *The voltages applied to the inverting and noninverting inputs must be equal.*

FIGURE 41-7 *Series regulator*

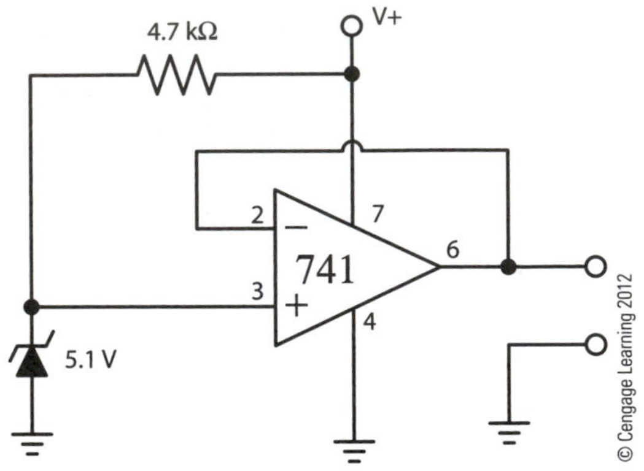

FIGURE 41-8 *Simple op amp regulator*

FIGURE 41-9 *Negative feedback can be used to change the output voltage.*

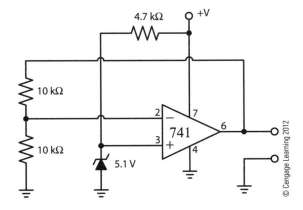

Testing the Op Amp

One of the easiest tests to perform on op amps in a working circuit is to measure the voltage drop between the inverting and noninverting inputs with a high-impedance voltmeter. If the voltage is not zero or close to zero, the op amp is probably defective. Therefore, if 5.1 volts is applied to the noninverting input, the inverting input must also be 5.1 volts. Because the inverting input is connected directly to the output, the output voltage must remain at 5.1 volts.

Changing the Output Voltage

The output voltage can be changed by connecting the inverting input to a voltage divider as shown in Figure 41-9. In this circuit, the inverting input is connected to a voltage divider made with two resistors of equal value. The inverting input must assume the same voltage as the noninverting input. Therefore, the output voltage of the op amp must become 10.2 volts to produce 5.1 volts at the inverting input. The output voltage of the regulator can be easily controlled by a simple voltage divider. If the voltage divider is replaced by a

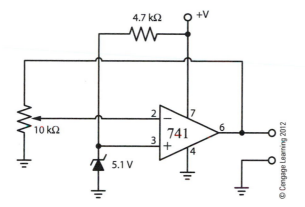

© Cengage Learning 2012

FIGURE 41-10 *Adjustable output voltage.*

potentiometer, as in Figure 41-10, the output voltage can be adjusted to any level between the reference voltage and the maximum output voltage of the op amp.

ADJUSTING THE REGULATOR

The regulator shown in Figure 41-10 still has a problem. If an adjustable regulator is needed, the output voltage must be adjustable to below 5.1 volts. This is easily done by using a very low-voltage zener diode that permits adjustment of the voltage to a low level. Remember that a common junction diode has a forward voltage drop of about 0.6 volt regardless of the current flow through it. The forward voltage drop of a junction diode, therefore, can be used as a zener of 0.6 volt, Figure 41-11. This circuit permits the output voltage to be adjusted to its lowest level (about 2 volts). Recall that when the V− connection at pin #4 is connected to ground instead of a voltage that is belowground, the output voltage cannot drop below about 2 volts.

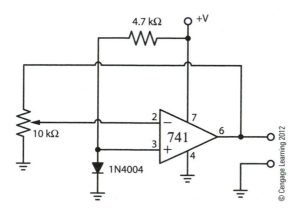

© Cengage Learning 2012

FIGURE 41-11 *Junction diode used as a reference*

FIGURE 41-12 *Belowground connection permits the output voltage to be adjusted to a low value.*

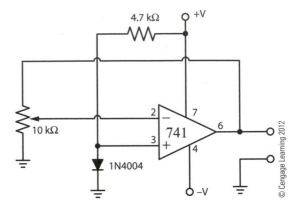

This condition can be corrected by connecting the op amp to an above- and belowground power supply, as shown in Figure 41-12. Connecting to a belowground voltage means that the output voltage can now be adjusted from 0.6 volt to the maximum the op amp can deliver.

Adding a Darlington Driver Circuit

The voltage regulator in Figure 41-12 has excellent regulation and is adjustable from 0.6 volt to the full output voltage of the op amp. It doesn't, however, have the ability to produce much current. The 741 operational amplifier has a maximum output current of about 5 mA. This 5 mA can be used to control the base of a transistor connected in series with the load. To improve circuit operation, a Darlington driver circuit is generally used to provide more transistor gain, Figure 41-13.

FIGURE 41-13 *Darlington amplifier circuit used to drive a power transformer*

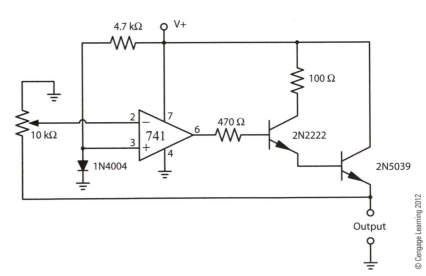

In this circuit, the potentiometer is connected to the emitter of the power transistor, not the output of the op amp. The connection is changed because the emitter of the transistor is now the output of the power supply and not the output of the op amp. The output voltage and current is limited only by the rating of the components.

CURRENT LIMIT

Many power supplies use a circuit that limits the output current to a safe level. If the output of the regulator in Figure 41-13 becomes shorted, enough current will flow to destroy the components in the circuit. Some power supplies have a current-limiting circuit. This permits adjustment of the maximum output current the regulator can produce. A simpler circuit can be used to limit the output current to a safe value. This is generally called short-circuit protection. Most current-limiting circuits used in DC power supplies operate by sensing the voltage drop across a low-value series resistor, Figure 41-14.

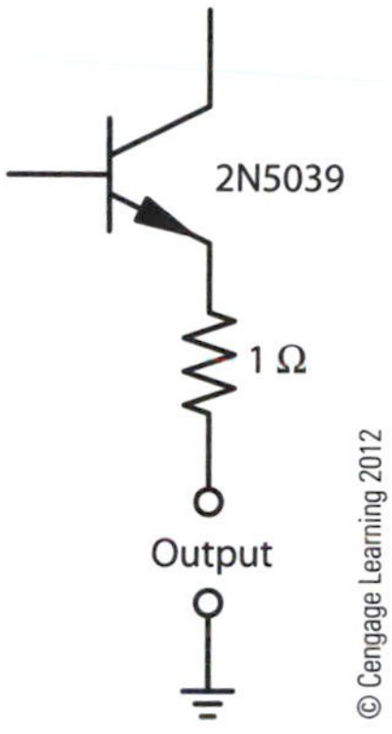

FIGURE 41-14 *One-ohm current-sensing resistor*

Voltage Drop

If a load is connected to the output terminals of the power supply, current flows through the series resistor. The voltage dropped across this resistor is proportional to the current flow. For instance, if 1 A of current flows through the resistor, a drop of 1 volt is developed across it. Because this drop is proportional to the current flow, a circuit can be built that does two things:

1. It senses the voltage drop.

2. It turns the transistor off when the current becomes excessive.

This same principle is used in small power supplies designed to operate electronic experiments to large SCR-controlled power supplies designed to provide the power needed to drive large DC motors. A common sensing resistor found in large power supplies designed to produce several hundred amperes is a precision $0.1\text{-}\Omega$ resistor rated at several hundred watts. In the power supply shown in Figure 41-15, a $1\text{-}\Omega$ wire wound resistor will be used.

Adding Short-Circuit Protection

The short-circuit protection circuit is very simple to construct, Figure 41-15. Transistors Q_1 and Q_2 form the Darlington driver pair that controls the output voltage. Transistor Q_3 provides short-circuit protection. When current flows through the $1\text{-}\Omega$ sensing resistor, a voltage drop is produced across it. Point A becomes more positive than point B. If Q_3 is a silicon PNP transistor, it begins to conduct

FIGURE 41-15 *Current limit circuit*

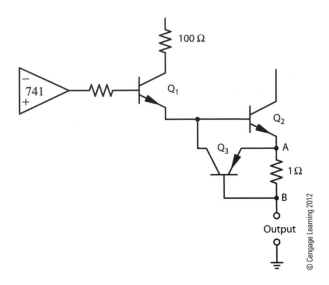

when the voltage between its emitter and base reaches about 0.6 volt. When Q_3 begins turning on, it steals the base current from Q_2 and causes it to begin turning off.

Note that transistor Q_3 will not permit a drop of more than about 0.6 volt to exist between its base and emitter connections. The current through the sense resistor does not become greater than that necessary to produce a drop of about 0.6 volt. In this circuit the current is limited to about 0.6 A ($1 \times 0.6 = 0.6$).

Changing the Current Limit

If a different current limit is desired, two changes can be made. One is to change the value of the sense resistor. For example, if another 1-Ω resistor is connected in parallel with the existing one, as shown in Figure 41-16, the total value of resistance is 0.5 Ω ($1/2 = 0.5$). This circuit requires a flow of about 1.2 A to produce a drop of 0.6 volt across the emitter–base junction of Q_3.

Another method is to increase the voltage drop needed to turn transistor Q_3 on. This is done by connecting a junction diode in series with the base of transistor Q_3, Figure 41-17. A diode connected in the forward bias direction has a voltage drop of 0.6 volt. Therefore:

- 0.6 volt drop of the emitter–base junction of Q_3 plus
- 0.6 volt drop of diode D_1 requires a drop across the sense resistor of
- 1.2 volts to turn Q_3 on.

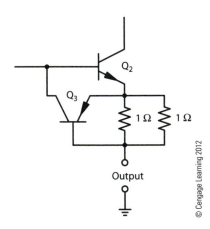

FIGURE 41-16 *Current-sensing resistor is decreased to 1/2 Ω.*

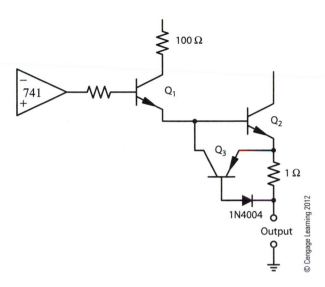

© Cengage Learning 2012

FIGURE 41-17 *Voltage drop of the current-limiting circuit is increased to about 1.2 volts.*

Of the two methods just described, the best is to lower the value of the sense resistor. This method places less resistance in series with the output of the power supply.

UNIT 41 REVIEW QUESTIONS

1. Why is there a drop in voltage at the output terminals of a power supply when a load is added?
   _____

2. A zener diode has dynamic impedance. What does this mean? _____
   _____

3. What is meant by the term *series regulator*? _____
   _____

4. What is meant by the term *shunt regulator*? _____
   _____

5. Name a disadvantage of a shunt regulator. _____
   _____

6. What determines the percentage of regulation for a series regulator? _____
   _____

7. How do most current-limiting circuits operate? _____
   _____
   _____

UNIT 42

NUMBER SYSTEMS

OBJECTIVES

After studying this unit, the student should be able to:

- Discuss binary number systems.

- Convert a binary number into a decimal number.

- Convert a decimal number into a binary number.

- Discuss the octal number system.

- Convert an octal number into a decimal number.

- Convert a decimal number into a binary number.

- Discuss the hexadecimal number system.

- Convert a hexadecimal number into a decimal number.

- Convert a decimal number into a hexadecimal number.

The base 10, or decimal, system is the most common of all number systems in use, but it is not the only one. The base 10 system employs 10 digits, 0 through 9, to represent all the values in the system. Computers and computer based devices such as programmable logic controllers often use *binary*, *octal*, or *hexadecimal* number systems. The logic gates discussed in the next unit are based on the binary number system.

THE BINARY NUMBER SYSTEM

The binary number system is a base 2 system in that it uses only two digits, 0 and 1, to represent all numerical values. Computers operate in the binary system because the 0s and 1s can represent on or off, true or false, or yes or no. Because different number systems do exist, subscripts are sometimes used to notate which system is being employed. The number (10_{10}) indicates the number 10 in the decimal system. The number (10_2) actually indicates the number 2 in the binary system.

Counting in Binary

Any number system is used to represent a certain quantity of items. In the illustration shown in Figure 42-1, a number of hexagon shaped nuts are used to illustrate both the decimal and binary number system.

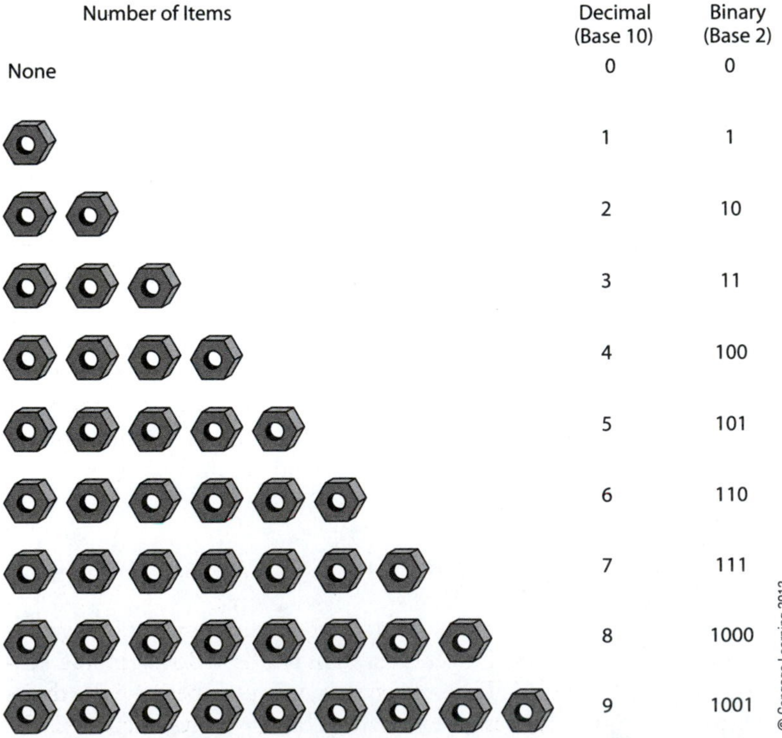

FIGURE 42-1 *Binary numbers that range from 0 through 9*

Number of Items	Decimal (Base 10)	Binary (Base 2)
None	0	0
	1	1
	2	10
	3	11
	4	100
	5	101
	6	110
	7	111
	8	1000
	9	1001

© Cengage Learning 2012

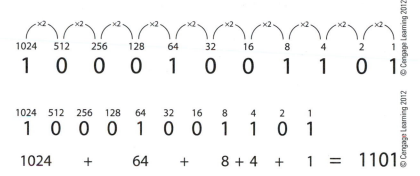

© Cengage Learning 2012

FIGURE 42-2 *Converting a binary number into a decimal number*

FIGURE 42-3 *Determining the equivalent decimal number*

© Cengage Learning 2012

Note that in the binary system only the numbers 0 and 1 are used. The numbers 2, 3, 4, 5, 6, 7, 8, and 9 do not exist.

Converting Binary to Decimal

When working with computers or programmable logic controllers, it may be necessary to convert a binary number into a common decimal number. This can be done by starting at the right-hand side of the binary number, with the number 1, and multiplying each number place to the left by 2 (because binary is a base 2 system). For example, convert the binary number 10001001101 to decimal, Figure 42-2.

After the appropriate decimal numbers have been placed above the binary numbers, bring down the decimal numbers that are located above a 1 and add them together, Figure 42-3.

The binary number 10001001101_2 is equivalent to 1101_{10} in decimal.

Converting Decimal to Binary

Decimal numbers can be converted to an equivalent binary number in a similar manner. The difference is that instead of multiplying by 2, you divide by 2. In this example, the decimal number 123_{10} will be converted to its binary equivalent. To perform this operation, divide the number by 2, Figure 42-4. If there is a remainder, place a 1 in the binary number position and subtract it from the quotient. If there is no remainder, place a 0 in the binary position and divide the quotient by 2 again. Repeat this procedure until the quotient is 0.

The equivalent binary number is 1111011_2.

THE OCTAL SYSTEM

The octal system was used by many early computers and programmable logic controllers. The octal system is a base 8 system and employs digits 0 through 7. The octal system was used because the digits 0 through 7 can be represented by three binary digits, Figure 42-5.

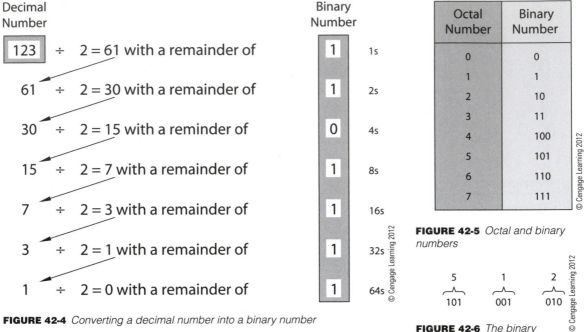

Decimal Number

123	÷	2 = 61	with a remainder of	
61	÷	2 = 30	with a remainder of	
30	÷	2 = 15	with a reminder of	
15	÷	2 = 7	with a remainder of	
7	÷	2 = 3	with a remainder of	
3	÷	2 = 1	with a remainder of	
1	÷	2 = 0	with a remainder of	

Binary Number

1	1s
1	2s
0	4s
1	8s
1	16s
1	32s
1	64s

FIGURE 42-4 *Converting a decimal number into a binary number*

Octal Number	Binary Number
0	0
1	1
2	10
3	11
4	100
5	101
6	110
7	111

FIGURE 42-5 *Octal and binary numbers*

5	1	2
101	001	010

FIGURE 42-6 *The binary equivalent of an octal number*

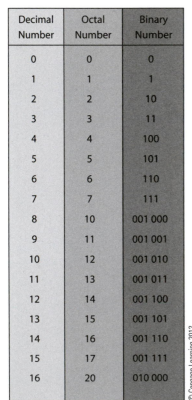

Decimal Number	Octal Number	Binary Number
0	0	0
1	1	1
2	2	10
3	3	11
4	4	100
5	5	101
6	6	110
7	7	111
8	10	001 000
9	11	001 001
10	12	001 010
11	13	001 011
12	14	001 100
13	15	001 101
14	16	001 110
15	17	001 111
16	20	010 000

FIGURE 42-7 *Decimal, octal, and binary numbers*

Unlike the binary system that employs long strings of binary digits to represent a number, the octal system employs binary digits in groups of three, Figure 42-6. The number 512_8 when converted to binary would be 101 001 010.

The chart shown in Figure 42-7 indicates equivalent decimal, octal, and binary numbers.

Converting Octal to Decimal

Octal numbers can be converted to decimal numbers in a similar manner to that of changing binary to decimal. When converting an octal number, each place to the left is multiplied by 8 instead of 2. Each place number is then multiplied by the octal number in that position, and the products are then added together.

EXAMPLE: Change 1262_8 to an equivalent decimal number. To perform this task, first start at the left and with the number 1 and multiply each place to the right by 8, Figure 42-8.

The next step is to multiply each of the place numbers by the octal number directly below the place number, Figure 42-9. The products are then added together. The octal number 1262_8 is equivalent to the decimal number 690_{10}.

Converting a Decimal Number to Octal

A decimal number can be changed to its octal equivalent in a similar manner to changing a decimal number to binary. The primary

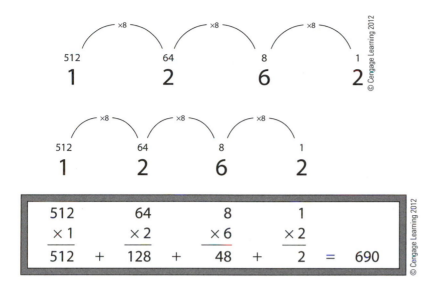

FIGURE 42-8 *Converting an octal number to a decimal number*

$$512 \quad\quad 64 \quad\quad 8 \quad\quad 1$$
$$1 \quad\quad\quad 2 \quad\quad\quad 6 \quad\quad\quad 2$$
(×8, ×8, ×8)

FIGURE 42-9 *Determining the decimal equivalent of an octal number*

$$512 \quad\quad 64 \quad\quad 8 \quad\quad 1$$
$$1 \quad\quad\quad 2 \quad\quad\quad 6 \quad\quad\quad 2$$
(×8, ×8, ×8)

512		64		8		1		
× 1		× 2		× 6		× 2		
512	+	128	+	48	+	2	=	690

FIGURE 42-10 *Converting a decimal number to an octal number*

$756_{10} \div 8 = 94$ with a remainder of   4

$94 \div 8 = 11$ with a remainder of   6

$11 \div 8 = 1$ with a remainder of   3

$1 \div 8 = 0$ with a remainder of   1

$$756_{10} = 1364_8$$

difference is that each quotient is divided by 8 instead of 2. The remainders are used to form the octal number.

EXAMPLE: Change the decimal number 756_{10} to its octal equivalent, Figure 42-10.

THE HEXADECIMAL NUMBER SYSTEM

The hexadecimal number system is used by most computers and programmable logic controllers. The reason is speed of operation. The hexadecimal system is a base 16 system and employs 16 characters to represent quantities. Because the decimal, or base 10, system contains only 10 characters used to indicate quantities (0 through 9), the hexadecimal system uses the alphabet letters A, B, C, D, E, and F to represent number values. Like the octal system, the hexadecimal system uses binary numbers in groups to represent an equivalent hexadecimal number. All the numbers in the hexadecimal system can be represented by a group of 4 binary digits. A chart showing decimal, hexadecimal, and binary numbers is shown in Figure 42-11.

Decimal Number	Hexadecimal Number	Binary Number
0	0	0000
1	1	0001
2	2	0010
3	3	0011
4	4	0100
5	5	0101
6	6	0110
7	7	0111
8	8	1000
9	9	1001
10	A	1010
11	B	1011
12	C	1100
13	D	1101
14	E	1110
15	F	1111

FIGURE 42-11 *Decimal, hexadecimal, and binary numbers*

© Cengage Learning 2012

Hexadecimal to Binary

	D	4
	1101	0100

Binary to Hexadecimal

	1111	0011
	F	3

© Cengage Learning 2012

FIGURE 42-12 *Converting hexadecimal and binary numbers*

Converting Hexadecimal and Binary Numbers

Hexadecimal numbers are converted to their binary equivalent by placing the binary numbers in groups of four digits. Binary numbers can be changed to their hexadecimal equivalents by the reverse process, Figure 42-12.

Converting Hexadecimal to Digital

Converting a hexadecimal number to its equivalent digital number is a very similar process to converting octal numbers to decimal. The difference is that each place number to the left is multiplied by 16 instead of 8. The place number is then multiplied by the hexadecimal number in that position. The products are then added together.

EXAMPLE: Change the hexadecimal number $3FA_{16}$ to an equivalent decimal number, Figure 42-13.

Converting Decimal to Hexadecimal

A decimal number can be converted to a hexadecimal number in the same basic manner that was used to change a decimal number to octal. In this instance, however, the decimal number is divided by 16 each time instead of 8, Figure 42-14.

EXAMPLE: Change 1256_{10} to an equivalent hexadecimal number.

FIGURE 42-13 *Converting a hexadecimal number into a decimal number*

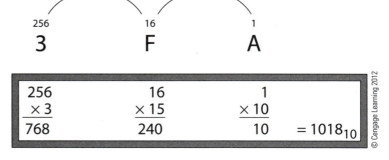

×16		×16	
256		16	1
3		F	A

256	16	1	
× 3	× 15	× 10	
768	240	10	= 1018_{10}

© Cengage Learning 2012

FIGURE 42-14 *Converting a decimal number to a hexadecimal number*

$1256_{10} \div 16 = 78$ with a remainder of   8

$78 \div 16 = 4$ with a remainder of (14)   E

$4 \div 16 = 0$ with a remainder of   4

$$1256_{10} = 4E8_{16}$$

© Cengage Learning 2012

UNIT 42 REVIEW QUESTIONS

1. A computer display turns lights on and off to represent binary numbers. A light turned on represents a high state, or 1. A light turned off represents a low state, or 0. What digital (base 10) number is represented by the display shown in Figure 42-15? _____

_____

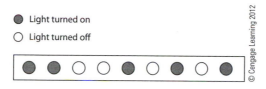

● Light turned on
○ Light turned off

FIGURE 42-15 *Computer display*

2. A programmable logic controller operates in the octal system. All numbers entered into the memory of the PLC must be entered in octal. The decimal number 89_{10} must be entered. What octal number should be entered into the PLC? _____

_____

3. What hexadecimal number would be represented by the lights on the computer display shown in Figure 42-16? Remember that binary numbers used to represent hexadecimal numbers are divided into groups of four. _____

_____

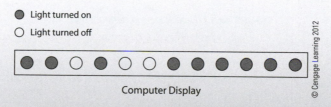

● Light turned on
○ Light turned off

Computer Display

FIGURE 42-16 *Determining the hexadecimal number*

4. A numerical machine is programmed by turning switches on or off to represent binary numbers. A switch turned on represents 1 and a switch turned off represents 0. The machine operates in the octal system. What binary number should be programmed for the octal number 742_8? _____

_____

UNIT 42 REVIEW QUESTIONS

5. Convert the decimal number 869_{10} into equivalent binary, octal, and hexadecimal numbers.

Binary _____

Octal _____

Hexadecimal _____

6. Convert the hexadecimal number $A3E_{16}$ into an equivalent octal number. _____
_____

7. A programmable logic controller operates in hexadecimal. The display indicates a value of F_3. What is this number equivalent to in decimal? _____
_____

8. The lights on a computer display shown in Figure 42-17 indicate a binary number. The computer operates in the octal system. What is the equivalent decimal number? _____
_____

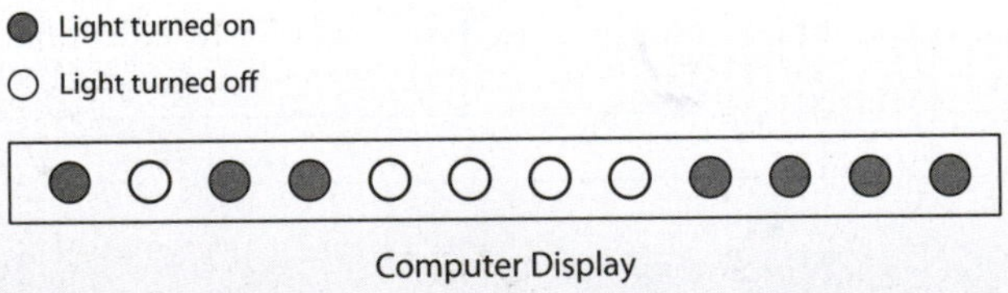

● Light turned on

○ Light turned off

● ○ ● ● ○ ○ ○ ○ ● ● ● ●

Computer Display

© Cengage Learning 2012

FIGURE 42-17 *Determining a decimal number*

9. If the computer in question 8 operated in hexadecimal instead of octal, what would be the hexadecimal number indicated by the display lights? _____
_____

10. What is the decimal equivalent of the hexadecimal number in question 9? _____
_____

UNIT 43

DIGITAL LOGIC

OBJECTIVES

After studying this unit, the student should be able to:

- List different types of logic.

- Discuss the operation of AND, OR, NAND, NOR, and INVERTER gates.

- Draw both USASI and NEMA logic symbols.

- Discuss the use of truth tables for different types of logic gates.

- Connect logic gates in a circuit and produce a truth table for each type of gate.

The electrician in industry today must be familiar with solid-state digital logic circuits. An example of these digital logic, or digital integrated, circuits is shown in Figure 43-1, attached to a heat sink. A digital device is one that has only two states, on or off. Most electricians have been using them for many years without realizing it. Magnetic relays, for instance, are digital devices. The coil is the input and the contacts are the output. They are considered to be single-input, multioutput devices, Figure 43-2.

Although relays are digital devices, the term *digital logic* has come to mean circuits that use solid-state control devices known as gates. The five basic types of gates are: the AND, OR, NOR, NAND, and INVERTER, or NOT. Each of these gates will be covered later in the unit.

VARIOUS TYPES OF DIGITAL LOGIC

There are also different types of logic. One of the earliest to appear was RTL, or resistor transistor logic. Next was DTL, or diode transistor logic, and then TTL, or transistor transistor logic. RTL and DTL have about faded out of existence, but TTL is still used to a fairly large extent. TTL can be identified because it operates on 5 volts.

FIGURE 43-2 *Magnetic relay*

FIGURE 43-1 *Inline integrated circuits*

Another type of logic frequently used in industry is HTL, which stands for high transit logic. It is better than TTL at ignoring the voltage spikes and drops caused by the starting and stopping of inductive devices such as motors. HTL generally operates on 15 volts.

Another very popular type of logic is CMOS, which has very high input impedance. CMOS is shortened from COSMOS, which means complementary symmetry metal oxide semiconductor. CMOS logic has the advantage of requiring almost no power to operate. There are also disadvantages.

- CMOS logic is very sensitive to voltage.

- The static charge in a body can sometimes destroy an integrated circuit (IC) just by touching the circuit.

People who work with CMOS logic often use a ground strap that is worn around the wrist like a bracelet. The strap is used to prevent a static charge from building up on the body.

A characteristic of CMOS logic is that unused inputs cannot be left in an indeterminent state. They must be connected to either a high or a low state.

THE *AND* GATE

Whereas magnetic relays are single-input, multioutput devices, gate circuits are multiinput, single output. For instance, an AND gate has at least two inputs, but only one output. Figure 43-3 shows an AND gate with three inputs, labeled A, B, and C, and one output labeled Y.

The AND gate symbol shown in Figure 43-3 is a USASI standard logic symbol. These symbols are commonly referred to as computer logic symbols. There is another system known as NEMA logic, which uses completely different symbols. The NEMA symbol for a three-input AND gate is shown in Figure 43-4.

Both symbols mean the same although they are drawn differently. Electricians in industry must learn both sets, because both types are used. Regardless of which symbol is used, the AND gate operates the same way.

The rule for an AND gate is that it must have all of its inputs high in order to get an output. Assuming that TTL logic is being used, a high level is considered to be +5 volts and low to be 0 volts.

The truth table shown in Figure 43-5 illustrates the state of a gate's output with different conditions of input. Ones are used to represent a high state and zeros a low state. Note in the figure that the output of the AND gate is high only when both of its inputs are high. The AND gate operates much like the simple switch circuit shown in Figure 43-6.

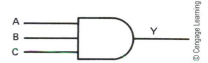

FIGURE 43-3 *USASI symbol of a three-input AND gate*

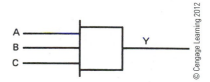

FIGURE 43-4 *NEMA logic symbol of a three-input AND gate*

A	B	Y
0	0	0
0	1	0
1	0	0
1	1	1

FIGURE 43-5 *Truth table for a two-input AND gate*

© Cengage Learning 2012

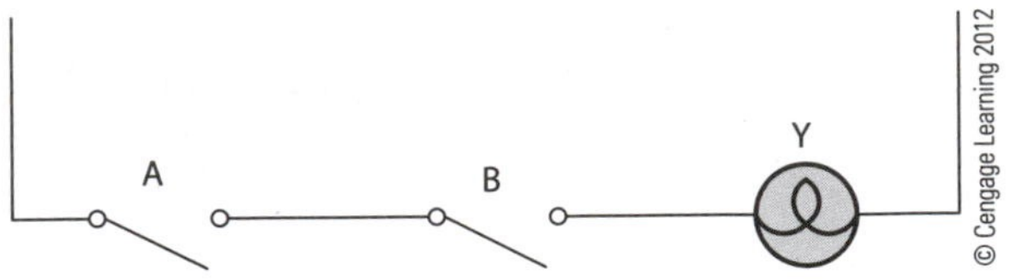

FIGURE 43-6 *Equivalent switch circuit of a two-input AND gate*

FIGURE 43-7 *Truth table for a three-input AND gate*

A	B	C	Y
0	0	0	0
0	0	1	0
0	1	0	0
0	1	1	0
1	0	0	0
1	0	1	0
1	1	0	0
1	1	1	1

FIGURE 43-8 *Equivalent switch circuit of a three-input AND gate*

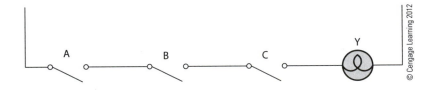

If a lamp is used to indicate the output of the AND gate, both A and B must be energized (switch position changed) before there is an output. Figure 43-7 shows the truth table for a three-input AND gate. There is still only one condition that permits a high output for the gate: when all inputs are high (at a logic level 1). When using an AND gate, remember: *any 0 input = a 0 output*. An equivalent switch circuit for a three-input AND gate is shown in Figure 43-8.

THE *OR* GATE

The next gate to study is the OR gate. Both the computer (USASI) and the NEMA logic symbols are shown in Figure 43-9.

The OR gate has a high output when either or both inputs are high: *any 1 input = a 1 output*. The truth table is shown in Figure 43-10. An equivalent switch circuit for the OR gate is shown in Figure 43-11. Note in this circuit that if either or both of the switches are energized, there will be an output at Y.

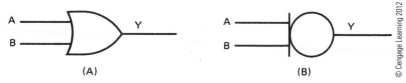

FIGURE 43-9 *(A) Computer logic symbol for an OR gate (B) NEMA logic symbol for an OR gate*

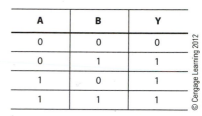

A	B	Y
0	0	0
0	1	1
1	0	1
1	1	1

FIGURE 43-10 *Truth table for a two-input OR gate*

THE *EXCLUSIVE-OR* GATE

Another gate that is very similar to an OR gate is known as an EXCLUSIVE-OR gate. The symbol for this gate is shown in Figure 43-12.

The EXCLUSIVE-OR gate has a high output when either, but not both, inputs are high. Refer to the truth table for this gate in Figure 43-13. Note that the equivalent switch circuit in Figure 43-14 shows that if both are de-energized or if both are energized, there is no output.

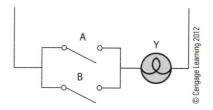

FIGURE 43-11 *Equivalent switch circuit of a two-input OR gate*

THE *INVERTER* (NOT) GATE

The simplest gate is the INVERTER, or NOT, gate. The INVERTER has one input and one output. As the name implies, the output is inverted (opposite the input). For example, if the input is high, the output will be low; and if the input is low, the output will be high. Figure 43-15 shows the computer and NEMA symbols for the INVERTER. Note that the computer symbol is an amplifier with a small circle (O), drawn at the output point. When using computer logic, an O drawn on a gate means to invert. Because the O appears on the output end of the amplifier, it means that the output is inverted. NEMA symbols use an X to show this. The truth table

FIGURE 43-12 *Computer logic symbol of an EXCLUSIVE-OR gate*

A	B	Y
0	0	0
0	1	1
1	0	1
1	1	0

FIGURE 43-13 *Truth table for a two-input EXCLUSIVE-OR gate*

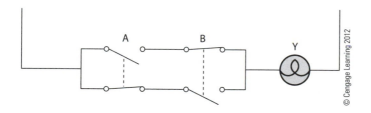

FIGURE 43-14 *Equivalent switch circuit of a two-input EXCLUSIVE-OR gate*

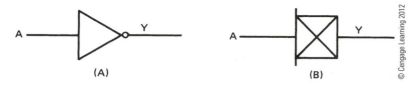

FIGURE 43-15 *(A) Computer logic symbol of an INVERTER (NOT). (B) NEMA logic symbol of an INVERTER (NOT)*

A	Y
0	1
1	0

© Cengage Learning 2012

FIGURE 43-16 *Truth table for an INVERTER (NOT)*

© Cengage Learning 2012

FIGURE 43-17 *Equivalent switch circuit of an INVERTER (NOT)*

in Figure 43-16 clearly shows that the output of the INVERTER is opposite the input. Figure 43-17 is an equivalent switch circuit for the INVERTER.

THE *NOR* GATE

The next gate to study is the NOR gate. The word *NOR* is shortened from NOT OR. The computer and NEMA logic symbols for a NOR gate are shown in Figure 43-18.

Note that the NOR gate symbol is the same as the OR gate symbol with an inverted output. A NOR gate is made by connecting an INVERTER to the output of an OR gate, as shown in Figure 43-19. The truth table in Figure 43-20 shows that the output of a NOR gate is 0 or low when any input is high: *any 1 input = a 0 output*. An equivalent switch circuit for this gate is shown in Figure 43-21. If either switch A or B is energized, there is no output at Y.

THE *NAND* GATE

The last gate to study is the NAND gate. NAND is shortened from NOT AND. Figure 43-22 shows the computer and the NEMA logic symbols for the NAND gate. These symbols are the same as the one for the AND gate with an inverted output. If any input of a NAND gate is low, the output will be high: *any 0 input = 1 output*. The output is 0 only when all inputs are 1. The truth table in Figure 43-23 clearly indicates these facts. Figure 43-24 shows an equivalent switch circuit

FIGURE 43-18 *(A) Computer logic symbol of a two-input NOR gate. (B) NEMA logic symbol of a two-input NOR gate*

FIGURE 43-19 *Equivalent NOR gate*

A	B	Y
0	0	1
0	1	0
1	0	0
1	1	0

© Cengage Learning 2012

FIGURE 43-20 *Truth table for a two-input NOR gate*

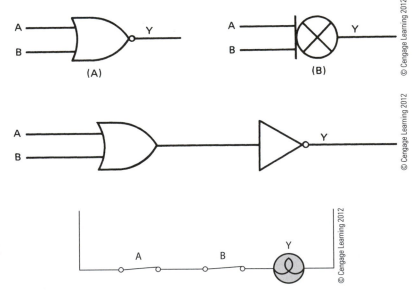

© Cengage Learning 2012

FIGURE 43-21 *Equivalent switch circuit of a two-input NOR gate*

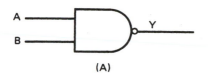

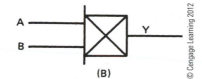

FIGURE 43-22 *(A) Computer logic symbol for a two-input NAND gate (B) NEMA logic symbol for a two-input NAND gate*

A	B	Y
0	0	1
0	1	1
1	0	1
1	1	0

FIGURE 43-23 *Truth table of a two-input NAND gate*

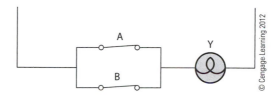

FIGURE 43-24 *Equivalent switch circuit of a two-input NAND gate*

FIGURE 43-25 *NAND gate connected as an INVERTER (NOT)*

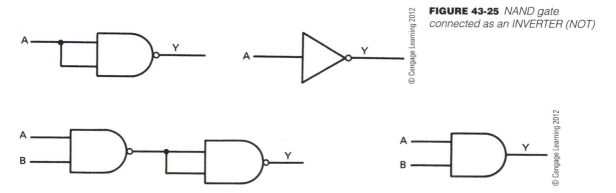

FIGURE 43-26 *NAND gates connected as an AND gate*

for the NAND gate. If either switch A or B is de-energized, there will be an output at Y.

The NAND gate is referred to as the *basic* gate because it can be used to make any of the other gates. Figure 43-25 shows the NAND gate connected to make an INVERTER. If a NAND gate is used as an INVERTER and connected to the output of another NAND gate, it becomes an AND gate, as shown in Figure 43-26.

VARIOUS GATE COMBINATIONS

When two NAND gates are connected as INVERTERS, and these INVERTERS are connected to the inputs of another NAND gate, the OR gate is formed, Figure 43-27.

If an INVERTER is added to the output of the OR gate shown in Figure 43-27, a NOR gate is formed, as in Figure 43-28.

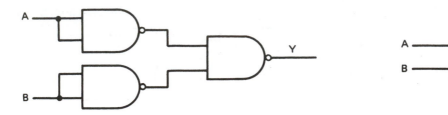

FIGURE 43-27 *NAND gates connected as an OR gate*

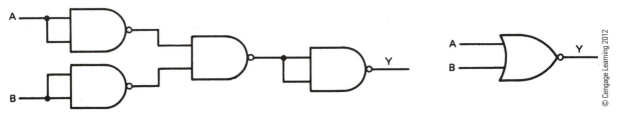

FIGURE 43-28 *NAND gates connected as a NOR gate*

HOUSING DIGITAL LOGIC GATES

Digital logic gates are generally housed in 14-pin IC packages. One of the types of TTL logic frequently used is the 7400 family of devices. For instance, a 7400 IC is a quad two-input positive NAND gate. This means that

- there are four NAND gates contained in the package;

- each gate has two inputs; and

- a level 1 is considered to be a positive voltage.

There can, however, be a difference in the way some ICs are connected. A 7400 (J or N) IC has a different pin connection than a 7400 (W) package. Note in Figure 43-29 that both ICs contain 4 two-input NAND gates, but the pin connections are different.

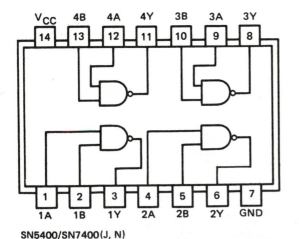

SN5400/SN7400(J, N)

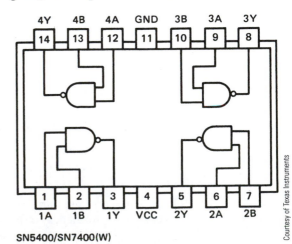

SN5400/SN7400(W)

FIGURE 43-29 *IC connection of a quad two-input NAND gate*

UNIT 43 REVIEW QUESTIONS

1. What do the letters RTL, DTL, and TTL stand for? _____

_____

_____

2. What type of logic has the highest input impedance? _____

_____

3. When using an AND gate, what must be done to have a high output? _____

_____

4. When using an OR gate, what must be done to have a high output? _____

_____

5. When using a NAND gate, what must be done to have a high output? _____

_____

6. When using a NOR gate, what must be done to have a high output? _____

_____

7. What is a digital device? _____

8. What is the basic difference between relay logic and gate logic? _____

_____

_____

9. Explain the operation of an EXCLUSIVE-OR gate. _____

_____

10. When using USASI logic symbols, what does a circle drawn on the output of a gate mean?

_____

_____

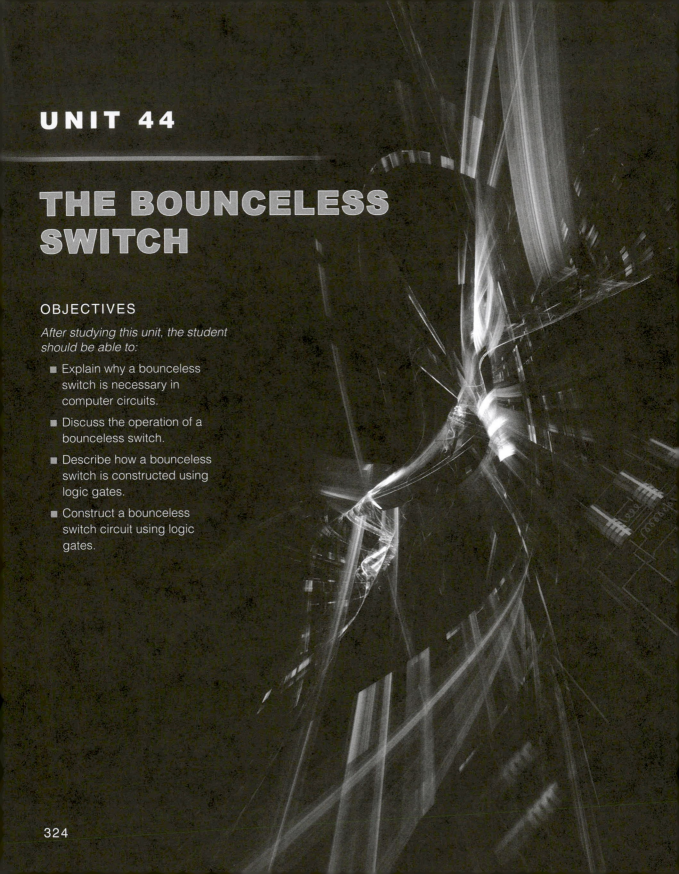

UNIT 44

THE BOUNCELESS SWITCH

OBJECTIVES

After studying this unit, the student should be able to:

- Explain why a bounceless switch is necessary in computer circuits.

- Discuss the operation of a bounceless switch.

- Describe how a bounceless switch is constructed using logic gates.

- Construct a bounceless switch circuit using logic gates.

When a control circuit is constructed, it must have sensing devices that tell it what to do. The number and types of devices used are determined by the circuit. They can range from a simple push button to float, limit, and pressure switches. Most of these devices use some type of mechanical switch to indicate their condition.

A float switch, for example, indicates its condition by opening or closing a set of contacts, Figure 44-1. The float switch can *tell* the control circuit whether a liquid is at a certain level or not. Most of the other sensing devices use this same method to indicate a condition. A pressure switch indicates that a pressure is at a certain level. Limit switches indicate whether a device has moved a certain distance or whether it is present or absent from some location.

FIGURE 44-1 *(A) Normally open float switch (B) Normally closed float switch*

THE SNAP-ACTION SWITCH

Almost all of these devices employ a *snap-action* switch. The snap action in a mechanical switch is generally obtained by spring-loading the contacts. The snap action ensures good contact when the switch operates.

Assume that a float switch is used to sense when water reaches a certain level in a tank. If the water rises at a slow rate, the contacts come together at a slow rate. This results in a poor connection. If the contacts are spring-loaded, however, they will snap from one position to another when the water reaches a certain level.

Contact Bounce

Most contacts have a snap action but they do not usually close with a single motion. When the movable contacts *make* with the stationary contact, there is often a fast, bouncing action. This means that the contacts may actually make and break three or four times in succession before the switch remains closed. When this switch controls a relay, contact bounce does not cause a problem. This is because relays are relatively slow-acting devices and are not affected, Figure 44-2.

When used with electronic control systems, contact bounce can be the source of a great deal of trouble. Most digital logic circuits are very fast acting. They can count each pulse when a contact bounces. Depending on the circuit, it may interpret each of these pulses as a

FIGURE 44-2 *Contact bounce does not greatly affect relay circuits.*

Relay

command. Contact bounce can cause the control circuit to literally "lose its mind."

Debouncing the Switch

For this reason, contacts are debounced before they are permitted to talk to the control system. When contacts must be debounced, a circuit commonly called a **bounceless switch** is used. Several circuits can be used to construct a bounceless switch, but the most common method uses digital logic gates. Any of the inverting gates can be used, but in this unit only two will be used.

OPERATION OF A BOUNCELESS SWITCH

Before construction of the circuit begins, the operation of a bounceless switch should first be discussed. The idea is to construct a circuit that locks its output either high or low when it detects the first pulse from the mechanical switch. If the output is locked in any position, it ignores any other pulses it may receive from the switch. The output of the bounceless switch is connected to the input of the digital control circuit. The control circuit receives only one instead of a series of pulses.

CONSTRUCTING THE CIRCUIT

The INVERTER

The first gate to be used to construct a bounceless switch is the INVERTER. The computer symbol and the truth table for this gate are shown in Figure 44-3. The bounceless switch circuit using INVERTERS is shown in Figure 44-4. The output of the circuit is high with the switch in the position shown. The switch connects the input of INVERTER #1 directly to ground or low. This causes the output of INVERTER #1 to be at a high state. This output is connected to the input of INVERTER #2. Because the input of INVERTER #2 is high, its output is low. The output of INVERTER #2 is connected to the input of INVERTER #1. This causes a low condition at the input of INVERTER #1.

If the position of the switch is changed, as shown in Figure 44-5, the output will change to low. The switch now connects the input

FIGURE 44-3 *(A) INVERTER symbol (B) INVERTER truth table*

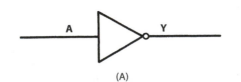

A	Y
0	1
1	0

(A)   (B)

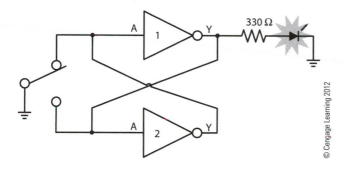

FIGURE 44-4 *High-output condition*

330 Ω

A 1 Y

A 2 Y

© Cengage Learning 2012

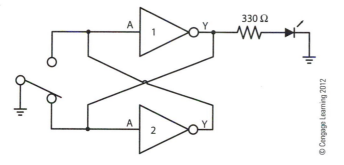

FIGURE 44-5 *Low-output condition*

330 Ω

A 1 Y

A 2 Y

© Cengage Learning 2012

of INVERTER #2 to ground or low. The output is, therefore, high. The high output of INVERTER #2 is connected to the input of INVERTER #1. Because INVERTER #1 has a high connected to its input, its output becomes low. The output of INVERTER #1 is connected to the input of INVERTER #2. This forces a low input to be maintained at INVERTER #2. Note that the output of one INVERTER is used to lock the input of the other one.

The NAND Gate

The second logic gate to be used to construct a bounceless switch is the NAND gate. The computer logic symbol and the truth table for this gate are shown in Figure 44-6. Figure 44-7 shows the construction of a bounceless switch using NAND gates.

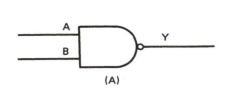

(A)

A	B	Y
0	0	1
0	1	1
1	0	1
1	1	0

(B)

© Cengage Learning 2012

FIGURE 44-6 *(A) NAND gate symbol (B) NAND gate truth table*

FIGURE 44-7 *High-output condition*

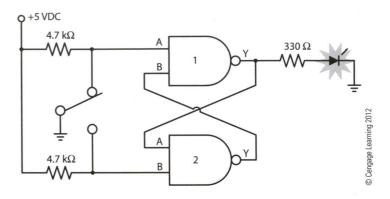

FIGURE 44-8 *Low-output condition*

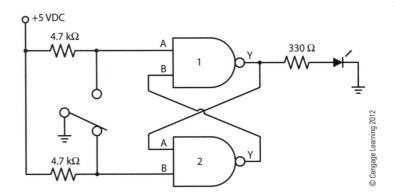

In this circuit, the switch has input A of gate #1 connected to low or ground. Because A is low, the output is high. The output of gate #1 is connected to input A of gate #2. Input B of gate #2 is connected to a high through the 4.7-kΩ resistor. Because gate #2 has both inputs high, its output is low. This output is connected to the B input of gate #1. Gate #1 now has a low connected to input B. Its output is forced to remain high even if contact bounce causes a momentary high at input A.

When the switch changes position, as shown in Figure 44-8, input B of gate #2 is connected low. This forces the output of gate #2 to become high. The high output is connected to the B input of gate #1. Input A of gate #1 is connected to high through a 4.7-kΩ resistor. Because both inputs of gate #1 are high, its output is low. The low is connected to the A input of gate #2. This forces its output to remain high even if contact bounce should cause a momentary high at the B input.

The output of this circuit remains constant even if the switch contacts should bounce. The switch is ready to be connected to the input of an electronic control circuit.

UNIT 44 REVIEW QUESTIONS

1. Why do most switches require a snap action? _____
_____

2. Why doesn't contact bounce cause a problem with relay circuits but causes trouble with computer-operated circuits? _____
_____

3. When a bounceless-switch circuit is constructed using logic gates, are the gates generally of the inverting or noninverting type? _____
_____

UNIT 45

DESIGN OF CIRCUIT #1
(Photodetector)

OBJECTIVES

After studying this unit, the student should be able to:

- Discuss the operation of solar cells.
- Discuss the operation of a cad cell.
- Discuss the design parameters of a photo circuit.
- Construct a light-sensitive circuit.

The first circuit to be designed is a photodetector. The circuit is used to turn on a light when darkness comes or activate a control component when a light is either present or absent. First, concentrate on turning on a light when it becomes dark.

One of the first things to be learned about circuit design is that there are generally several ways any particular circuit can be constructed and still perform the same function. The final design is usually determined by several factors, including the availability of components and their cost. Other facts can also be important. These include whether the circuit must be operated in extremes of temperature and whether the circuit must be battery powered.

REQUIREMENTS FOR THE CIRCUIT

One of the first steps in design is to determine what is needed to do the job. This circuit is to be used to turn a light on when it becomes dark and to turn the light off again when it becomes light. It must, therefore, have two basic items:

1. Some device to detect the presence or absence of light

2. A method to connect the light to the power line

Two components immediately come to mind: solar cells and a relay.

Solar Cells and the Relay

If enough solar cells are connected to operate a relay, the circuit in Figure 45-1 will fulfill the requirements. The light is connected in series with a normally closed relay contact to 120 volts AC. The contact is controlled by CR relay coil, and silicon solar cells are connected to the coil. During the daylight hours, the solar cells provide power to the relay coil, which keeps the contact open. Because the relay contact is held open during the daylight hours, the light is turned off. During the hours of darkness, the solar cells cannot produce the power needed

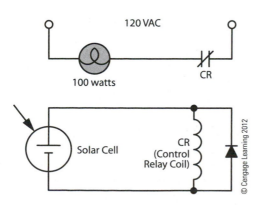

FIGURE 45-1 *Photodetector operated by solar cells*

to operate the relay coil, and the contact closes. When it closes, the light is connected to the 120-volt AC power line.

This circuit fulfills all the requirements. A different one may have to be used, however, because silicon solar cells are expensive. One cell produces only about 0.5 volt. If the relay coil operated on 12 volts, it would require 24 solar cells connected in series to produce the 12 volts needed to power the relay coil.

CAD CELLS: Another device that can be used to detect light is the cadmium sulfide cell, or **cad cell**, Figure 45-2. The cad cell does not produce a voltage in the presence of light as the solar cell does, but rather changes its resistance with a change of light. When it is exposed to a bright light, its resistance drops to a low value. When it is in darkness, its resistance increases greatly. The exact amount of light or dark resistance varies from one cad cell to another. Typically it has about 50 Ω in direct sunlight and several hundred thousand ohms in darkness. Cad cells are economical and can be obtained at almost any electronic supply store.

Connection #1: Examine the circuit shown in Figure 45-3. The light is connected in series with a normally closed relay contact. The coil of the 12-volt DC relay is connected in series with a cad cell to the DC power supply. When the cad cell is exposed to sunlight, its resistance drops low enough to permit the relay coil to turn on. When the coil turns on, the normally closed contact opens and disconnects the light from the power source. When the cad cell is in darkness, its

FIGURE 45-2 *Cad cells*

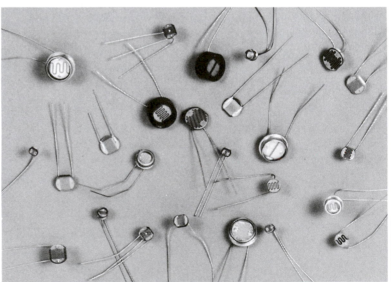

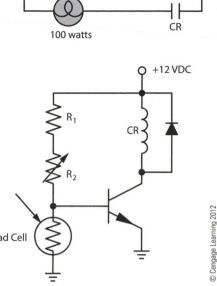

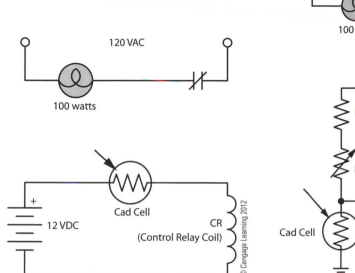

FIGURE 45-3 *Relay coil connected to a cad cell*

FIGURE 45-4 *Cad cell used to control a transistor*

resistance increases and turns the relay off. When the relay turns off, the contact closes and connects the light to the AC power source.

This circuit appears to perform all the requirements that were originally set for its operation. There are several problems with it, however. One is that cad cells are very low-power devices and not many are able to control the current through the relay without being destroyed. Another is the lack of control. There is no method that can be used to control when the light will turn on or off. A third is that the cad cell will not provide a definite on or off action for the relay. Its resistance changes with a change of light intensity. It does not operate like a digital device, which has either a high or a low resistance. Because the cad cell is not a digital device, the current slowly increases through the relay coil as the resistance of the cad cell decreases. This slow increase or decrease of current prevents the relay from operating properly.

Connection #2: The circuit shown in Figure 45-4 is an improvement over that in Figure 45-3. In this circuit, the light is connected in series with a normally open contact that is controlled by CR relay coil. The coil is connected in series with a transistor that is controlled by the cad cell.

When the cad cell is in the presence of light, its resistance is very low. In this condition, it steals the base current to ground and the

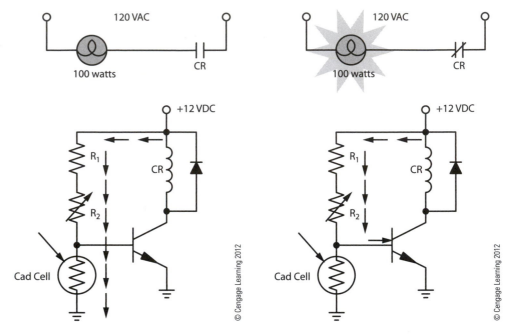

FIGURE 45-5 *Cad cell resistance is low in the presence of light.*

FIGURE 45-6 *Cad cell resistance is high in the presence of darkness.*

transistor is turned off, Figure 45-5. When the cad cell is in darkness, its resistance increases and current is permitted to flow to the base of the transistor, Figure 45-6. When current flows to the base of the transistor, the transistor turns on and energizes the relay. When the relay energizes, CR contacts close and connect the light to the 120-volt AC power source.

This circuit is a definite improvement over that shown in Figure 45-3. The transistor of this circuit can easily control the current needed to operate the relay. The variable resistor, R_2, can adjust the sensitivity of the circuit. Although this circuit has corrected some of the problems of the circuit in Figure 45-3, it still retains one of them.

The transistor is not a digital device and will not turn the coil on or off with a snap action. If this circuit is to operate properly, the photodetector should work with a definite on or off action. A device is required that can convert the analog operation of the cad cell into a digital action.

Adding an Op Amp

The operational amplifier can do this conversion when it is used as a level detector. The circuit in Figure 45-7 uses the cad cell to control

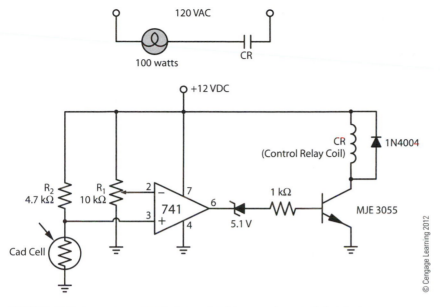

FIGURE 45-7 *An op amp permits the cad cell to operate as a digital device.*

the output of the operational amplifier. Resistor R_1 sets the inverting input to some positive voltage. Assume this voltage level to be 5 volts. Resistor R_2 and the cad cell form a voltage-divider circuit for the noninverting input. When the cad cell is in the presence of light, its resistance is low. The voltage applied to the noninverting input is less than 5 volts. Therefore, the output of the op amp is low. Because the output is low, the transistor is turned off and CR relay is de-energized. When the cad cell is in darkness, its resistance increases above 4.7 kΩ. The voltage applied to the noninverting input becomes greater than 5 volts. This causes the output of the op amp to change to a high state and supply base current to the transistor. The transistor turns the relay coil on and closes contact CR, which connects the light to the power line.

Note that a circuit now exists that will perform the desired job. It is inexpensive to construct, and the light sensitivity can be adjusted by resistor R_1. The op amp permits the cad cell to operate as a digital device. This means that the relay can be turned completely on or completely off.

CHANGING THE CIRCUIT

Although the circuit design goal has been completed, assume that a change in the sense of the photodetector is desired. In other words, the circuit as originally designed turns a light on when the cad cell is

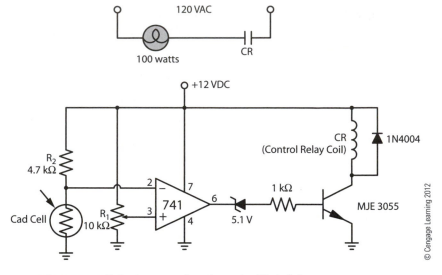

FIGURE 45-8 *Relay turns on when the cad cell is in light.*

in darkness and turns the light off when the cell is in the presence of light. The change involves altering the circuit so that the light will turn on when the cad cell is in the presence of light and off when it is in darkness, Figure 45-8.

Examine this circuit and note that it is basically the same as the circuit shown in Figure 45-7. The only differences are:

1. Resistor R_1 has been connected to the noninverting input.

2. The voltage divider, formed by the 4.7-kΩ resistor and the cad cell, has been connected to the inverting input.

Now assume that resistor R_1 has been adjusted to produce 5 volts at the noninverting input. When the cad cell is in the presence of light, its resistance is low. A voltage of less than 5 volts is applied to the inverting input. Therefore, the output of the op amp is high and CR relay is turned on. When the cad cell is in darkness, its resistance increases above 4.7 kΩ. This causes a voltage greater than 5 volts to be applied to the inverting input. The output of the op amp goes to a low state and turns off the CR relay. Note that this circuit causes the light to turn on when the cad cell is in the presence of light and off when it is in darkness. With a few simple changes to the original circuit, there is now a circuit with the opposite sense of operation.

UNIT 45 REVIEW QUESTIONS

1. Define the difference between a solar cell and a cad cell. _____
   _____
   _____

2. What is the average voltage produced by a solar cell in the presence of light? _____
   _____

3. What is the typical resistance of a cad cell in the presence of direct sunlight? _____
   _____

4. Why was an operational amplifier used in this circuit? _____
   _____

5. What change was made to permit the relay to turn on when the cad cell was in the presence of light instead of darkness? _____
   _____

UNIT 46

DESIGN OF CIRCUIT #2
(Animal Feeder)

OBJECTIVES

After studying this unit, the student should be able to:

- Discuss designing circuits in blocks and then putting the blocks together to form the required circuit.

- Construct a complex circuit by building the circuit in blocks and then connecting the blocks together.

CIRCUIT REQUIREMENTS

When a circuit that is large and complex is to be designed, it is generally designed in blocks. This circuit has been chosen to illustrate this principle. Originally, the circuit was designed as an automatic animal feeder, but with a few changes it could be used for many applications. The requirements for it are as follows:

1. It must be operated by battery so it can be used in remote locations.

2. It must operate twice each day.

3. Each time it activates, it will run for 3 seconds and turn off.

CIRCUIT EXPLANATION

A short explanation of the circuit will aid in the understanding of the requirements.

- This circuit is to be used to operate a small DC motor.
- The motor is connected to a metal disk.
- The disk has flat metal blades attached to it.
- Grain is placed into a hopper that is located above the disk.
- When the motor runs, the disk slings grain in all directions to feed the animals.

MEETING THE REQUIREMENTS

Requirement #1 can be met by using components that require only a small amount of power to operate. This circuit was designed to operate for at least 30 days on a 12-volt automobile battery with a capacity of about 60 Ah.

Requirement #2 is more difficult. When thinking of an operating device that functions twice each day, the time clock usually comes to mind. More specifically, a digital alarm clock is the first thought to meet this requirement. Unfortunately, a common digital alarm clock will not be useful. This circuit requires a device that will produce one pulse each time it alarms. This requirement can be met, then, by using the Sun as a clock and designing the circuit so that the motor will operate in the morning and again in the evening. Because the sun is to be used as the timer, some type of photodetector will be required to sense when the sun rises and sets. The photodetector that was designed in Unit 44 should be able to perform this job.

Resistor R_1 in Figure 46-1 is used to preset a voltage between +12 and 0 volt at the noninverting input. It adjusts the light sensitivity of the photodetector. The 4.7-kΩ resistor and the CAD cell

FIGURE 46-1 *Photodetector circuit*

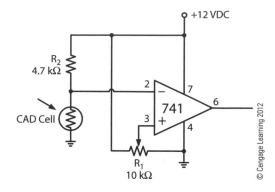

form a voltage divider that is connected to the inverting input. When the cad cell is in the presence of light, its resistance is low. This permits a low voltage to be present at the inverting input. This voltage should be less than the preset voltage applied to the noninverting input. Because the voltage applied to the noninverting input is at a higher level than that applied to the inverting input, the output voltage goes high.

In the presence of darkness, the resistance of the cad cell increases in value. This also causes the voltage applied to the inverting input to increase. When the voltage applied to the inverting input becomes higher than the voltage applied to the noninverting input, the output changes to a low state. Note that when the cad cell is in the presence of light, the output is high, and when in darkness, the output is low.

Requirement #3 states that the motor should operate for only three seconds each time it is activated. If this requirement is to be met, a timer must be used.

In the circuit shown in Figure 46-2, the photodetector has been coupled to a 555 timer. Recall that in the operation of the 555 timer, the trigger, pin #2, must be connected to a voltage less than one-third of Vcc to trigger the unit. When the timer is triggered, the output, pin #3, turns on. The discharge, pin #7, turns off. When the output turns on, transistor Q_1 turns on and energizes the CR relay. When the discharge turns off, the capacitor connected to the threshold, pin #6, begins to charge through resistors R_3 and R_4. When it has been charged to a voltage greater than two-thirds of Vcc, the output attempts to turn off and the discharge to turn on. This cannot happen, however, because the trigger is now connected to a voltage less than one-third of Vcc. The timer remains permanently triggered as long as the output of the op amp is low. The trigger must be *pulsed* if the timer is to operate as a monostable device.

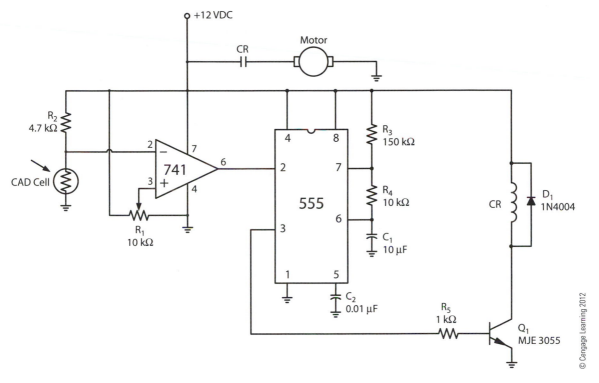

FIGURE 46-2 *Photodetector connected directly to a 555 timer circuit*

If this part of the circuit is to operate, the trigger must receive a low pulse when the op amp changes from a high to a low state. Adding the 0.1-μF capacitor and the 100-kΩ resistor, Figure 46-3, permits the trigger to receive a one-time pulse when the output of the op amp changes from a high to a low state. The 100-kΩ resistor, R_6, charges capacitor C_3 to 12 volts. This 12-volt charge keeps the trigger of the 555 timer connected to Vcc. When the output of the op amp changes from a high to a low state, capacitor C_3 is suddenly discharged. Resistor R_6 permits the capacitor to recharge and again supply 12 volts to the trigger of the timer.

If an oscilloscope is connected to the trigger pin of the timer, it indicates a value of 12 volts until the op amp changes to a low state. When the change takes place, the voltage will pulse low and then return to 12 volts. A waveform similar to the one in Figure 46-4 can be seen on the display of the oscilloscope.

This addition to the circuit permits the op amp to provide a single low pulse to the trigger of the timer each time the op amp changes from a high to a low state. When the trigger of the timer receives a low pulse from the op amp, the timer output turns on and

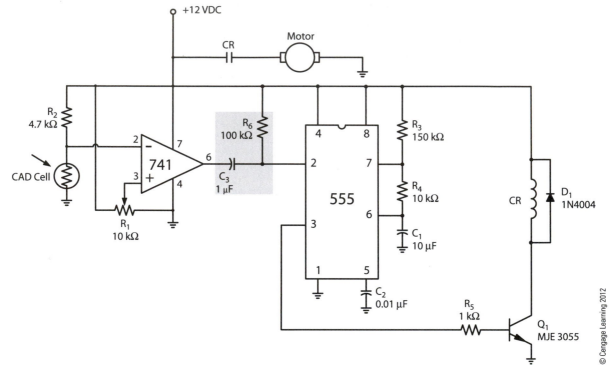

FIGURE 46-3 R_6 and C_3 have been added to produce a pulse.

FIGURE 46-4 Low pulse produced when the op amp changes to a low state

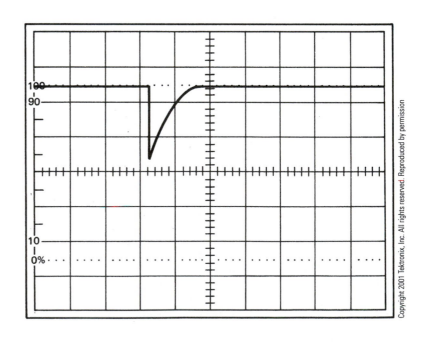

energizes the CR relay. When the capacitor that is connected to the threshold charges to two-thirds of Vcc, the output turns off and the discharge turns on. The timer remains in this state until the trigger receives another low pulse from the op amp.

This circuit will now operate the motor for approximately 3 seconds each time the op amp changes from a high to a low state or when daylight changes to dark. A requirement of the circuit, however, states that the timer must operate twice each day. A method must be found to trigger a timer when the photodetector changes from darkness to light.

When the op amp changes from darkness to light, the output changes from a low to a high state. The 555 timer, however, will not operate when triggered by a high pulse. Therefore, some method must be found to convert the high pulse of the op amp into a low pulse.

The addition of the components shown in Figure 46-5 should produce a low pulse at point B when the output of the op amp changes from a low to a high state. When the output of the op

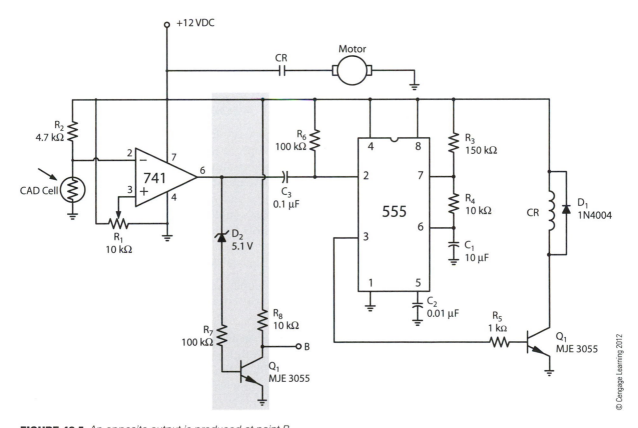

FIGURE 46-5 *An opposite output is produced at point B.*

amp is low, the voltage is less than the reverse breakdown voltage of the zener diode. No current can flow to the base of transistor Q_2. Because Q_2 is turned off, the output voltage at point B is 12 volts. When the output of the op amp changes from a low to a high state, the voltage becomes greater than the reverse breakdown voltage of the zener diode, and current flows to the base of transistor Q_2. When Q_2 turns on, the voltage at point B drops to about 0.7 volt. Note that the voltage at point B is opposite the output of the op amp. When the output of the op amp is low, the voltage at point B is high. When the output of the op amp is high, the voltage at point B will be low.

Because a low voltage is now produced at point B when the photodetector changes from darkness to light, it becomes a simple matter to connect another timer circuit. Note in Figure 46-6 that the circuit for timer B is the same as the circuit for timer A. Diodes D_3 and D_4 have been connected to the output terminals of the timers. The diodes form a simple OR circuit that permits either timer to turn on transistor Q_1 without causing feedback to the output of the other timer.

For example, if the output of timer A turns on, diode D_3 is forward biased and permits current flow to the base of transistor Q_1. When D_3 is conducting, D_4, however, is reverse biased and does not permit current to flow to the output of timer B.

SOLVING CIRCUIT PROBLEMS

The only problem that may occur with this circuit will probably be caused by the motor. Because it is a DC motor, it contains brushes and a commutator. As the brushes make and break contact with the commutator, voltage spikes are produced. These can be interpreted as commands by the timers and will sometimes cause the timers to retrigger. This in turn could cause the motor to operate several times in succession before it stops running. Note in Figure 46-6 that a 50-μF capacitor has been connected across the motor. It will greatly reduce any voltage spikes produced by the motor.

The circuit is now complete. All requirements for its operation have been met. Note that the circuit is composed mainly of a photodetector and two timers.

The only real design problem was how to connect the photodetector circuit to the two timer circuits. As circuits become larger and more complex, it will be found that they are designed by first constructing the major portions of the circuit as blocks and then connecting the blocks together.

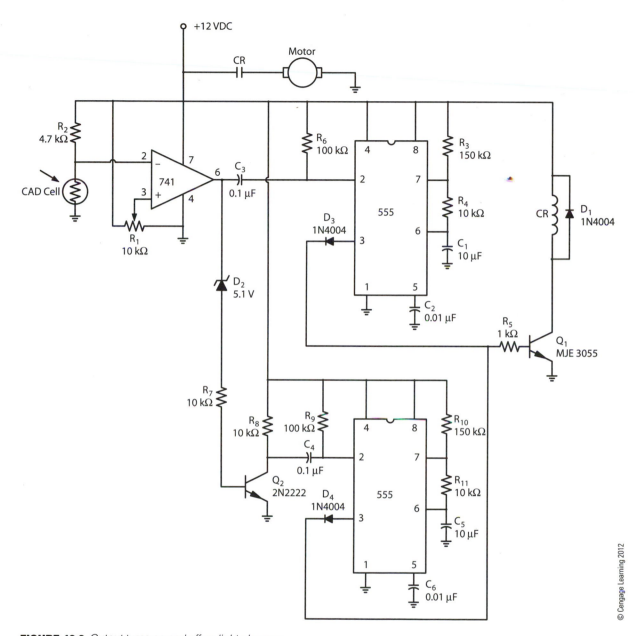

FIGURE 46-6 *Output turns on and off as light changes.*

UNIT 46 REVIEW QUESTIONS

To answer the following questions, refer to the circuit shown in Figure 46-6.

1. What is the function of resistor R_1? _____

_____

2. What is the function of transistor Q_2? _____

_____

3. What is the function of diode D_1? _____

_____

4. What determines the amount of time the output of timer B will be turned on after it has been

triggered? _____

_____

5. What is the function of capacitor C_3? _____

_____

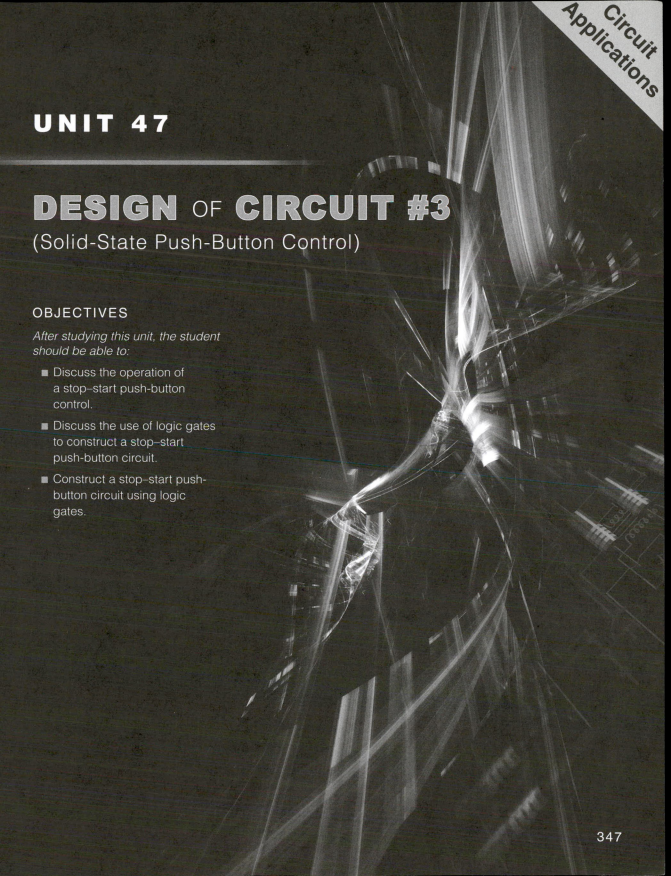

UNIT 47

DESIGN OF CIRCUIT #3
(Solid-State Push-Button Control)

OBJECTIVES

After studying this unit, the student should be able to:

- Discuss the operation of a stop–start push-button control.

- Discuss the use of logic gates to construct a stop–start push-button circuit.

- Construct a stop–start push-button circuit using logic gates.

In this unit, a digital circuit will be designed to perform the same function as a common relay circuit. The relay circuit to be used is a basic stop–start push-button circuit with overload protection, Figure 47-1.

THE COMMON RELAY CIRCUIT

Before beginning the design of an electronic circuit that performs the same function as the relay circuit, the operation of the relay circuit should be discussed.

In the circuit shown in Figure 47-1, no current can flow to relay coil M because of the normally open start button and the normally open contact controlled by relay coil M.

Circuit Operation

When the start button is pushed, current flows through the relay coil and normally closed overload contact to the power source, Figure 47-2. When current flows through the relay coil M, the contacts connected in parallel with the start button close. These contacts maintain the circuit to coil M when the start button is released and returns to its open position, Figure 47-3.

The circuit continues to operate until the stop button is pushed and breaks the circuit to the coil, Figure 47-4. When the current flow to the coil stops, the relay de-energizes and contact M reopens. Because the start button and contact M are open, there is no complete circuit to the relay coil when the stop button is returned to its normally closed position. To restart the relay, the start button

FIGURE 47-1 *Start–stop push-button circuit*

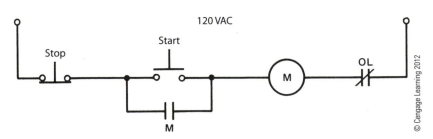

FIGURE 47-2 *Start button energizes M relay coil.*

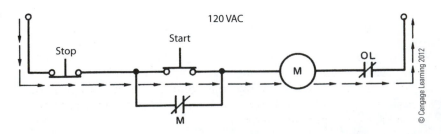

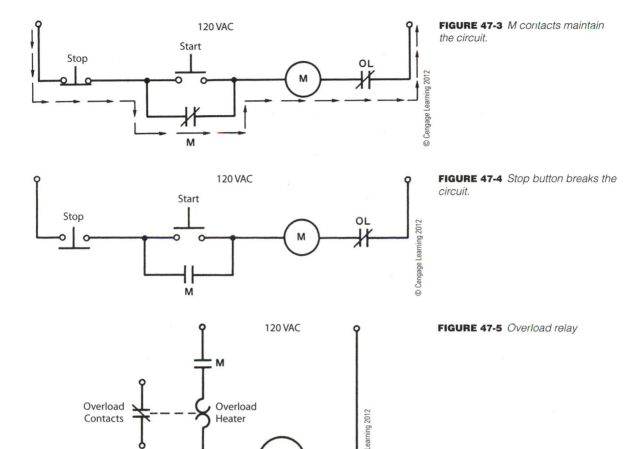

FIGURE 47-3 *M contacts maintain the circuit.*

FIGURE 47-4 *Stop button breaks the circuit.*

FIGURE 47-5 *Overload relay*

must be pushed again to provide a complete circuit to the coil of the relay.

The only other logic condition that can occur in this circuit will be caused by the motor connected to the load contacts of M relay. The motor is connected in series with the heater of an overload relay, Figure 47-5. When M coil energizes, it closes the load contact M shown in Figure 47-5. When this contact closes, it connects the motor to the 120-volt AC power line. If the motor becomes overloaded, it causes too much current to flow through the circuit. When a current greater than normal flows through the overload heater, it causes the heater to produce more heat than it does under normal conditions. If the current becomes high enough, it causes the normally closed overload contact to open. Note that this contact is electrically isolated from the heater. It can therefore be connected to a totally different voltage source than the motor.

FIGURE 47-6 *Overload contacts break the circuit.*

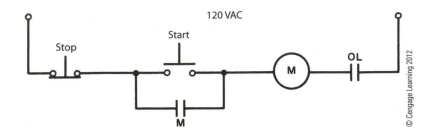

If the overload contact should open, the control circuit is broken and the relay de-energizes, Figure 47-6. The open contact affects the circuit in the same way as pushing the stop button. After it has been reset to its normally closed position, the coil remains de-energized until the start button is again pressed.

DESIGNING THE DIGITAL LOGIC CIRCUIT

Now that the logic of the circuit is understood, the design of a digital logic circuit that will operate in this manner can begin.

The first problem is to find a circuit that will turn on with one push button and off with another. The circuit shown in Figure 47-7 can do this.

Operation of the Circuit

This circuit consists of an OR and an AND gate, and functions as follows:

1. Input A of the OR gate is connected to a normally open push button, which is connected to +5 volts DC.

2. The B input of the OR gate is connected to the output of the AND gate.

3. The output of the OR gate is connected to the A input of the AND gate.

FIGURE 47-7 *Output turns on and off with push buttons.*

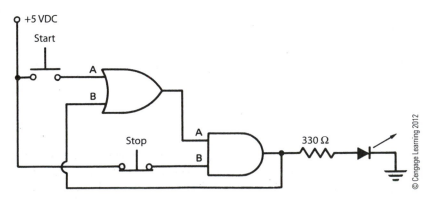

4. The B input of the AND gate is connected through a normally closed push button to +5 volts DC.

5. This push button is used as the stop button.

6. The output of the AND gate is the output of the circuit.

Understanding the Logic of the Circuit

To understand the logic of this circuit, assume that the output of the AND gate is low. This produces a low at input B of the OR gate. Because the push button connected to the A input is open, a low is also produced at this input. (Note: When all inputs of an OR gate are low, its output is also low.) The low output of the OR gate is connected to the A input of the AND gate. The B input of the AND gate is connected to a high through the normally closed push-button switch. Because the A input of the AND gate is low, its output is forced to remain in a low state.

When the start button is pushed, a high is connected to the A input of the OR gate. This causes the output of the OR gate to change to high. The high is connected to the A input of the AND gate. This gate now has both its inputs high, so its output changes from a low to a high state. When the output of the AND gate changes to a high state, the B input of the OR gate also becomes high. Because the OR gate now has a high connected to its B input, its output will remain high when the push button is returned to its open condition and input A becomes low. Note that this circuit operates the same way as the relay circuit when the start button is pushed. The output changes from a low to a high state and the circuit locks in this condition so the start button can be reopened.

When the normally closed stop button is pushed, the B input of the AND gate changes from a high to a low. When it changes to a low state, the output of the AND gate also changes to a low state. This causes a low to appear at the B input of the OR gate. This gate now has both inputs low, so its output changes from a high to a low state. Because the A input of the AND gate is now low, the output is forced to remain low when the stop button returns to its closed position and a high is connected to the B input. A circuit now exists that can be turned on with the start button and off with the stop button.

Adding the Overload Contact

The next design problem is to connect the overload contact into the circuit. It must be done in such a way that it causes the output of the circuit to turn off if the overload contact should open. The first impulse probably is to connect the circuit as shown in Figure 47-8.

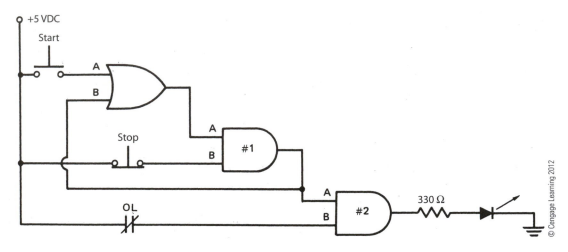

FIGURE 47-8 *Overload can turn off output.*

In this circuit, the output of AND gate #1 is connected to the A input of AND gate #2. The B input of AND gate #2 is connected to high through the normally closed overload contact. If the overload contact remains closed, the B input remains high. The output of AND gate #2 is therefore controlled by the A input. If the output of AND gate #1 changes to a high state, the output of AND gate #2 also changes to a high state. If the output of AND gate #1 becomes low, the output of AND gate #2 will become low too.

If the output of AND gate #2 is high and the overload contact opens, the B input becomes low and the output changes from a high to a low state. This circuit appears to operate with the same logic as the relay circuit until the logic is closely examined.

Assume that the overload contact is closed and the output of AND gate #1 is high. Because both inputs of AND gate #2 are high, the output is also high. Now assume that the overload contact opens and causes B input to change to a low condition. This forces the output of AND gate #2 also to change to a low state. The A input of AND gate #2 is still high, however. If the overload contact is reset, the output immediately changes back to a high state. If the overload contact opens and is then reset in the relay circuit, the relay will not restart itself. The start button has to be pushed to restart the circuit. Although this is a small difference in circuit logic, it could become a safety hazard in some cases.

CORRECTING FAULTS: DESIGN CHANGE

This fault can be corrected with a slight design change, Figure 47-9. In this circuit, the normally closed stop button is connected to the

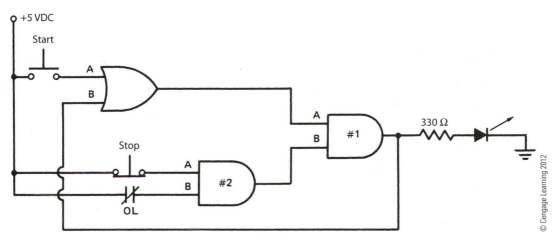

FIGURE 47-9 *Correct circuit operation*

A input of AND gate #2, and the normally closed overload switch is connected to the B input. As long as both of these inputs are high, the output of AND gate #2 provides a high to the B input of AND gate #1. If either the stop button or the overload contact opens, the output of AND gate #2 changes to a low state. When the B input of AND gate #2 changes to a low state, it causes the output of AND gate #1 to change to a low state and unlock the circuit just as pushing the stop button did in Figure 47-8. The logic of this digital circuit is now the same as the relay circuit.

Correcting Faults: Gate Connections

Although the logic of this circuit is now correct, there are still some problems. When the start button is in its normal position, the A input of the OR gate is not connected to anything. When an input is left in this condition, the gate may not be able to determine whether the input should be high or low. The gate could therefore assume either condition. To prevent this, gate inputs must always be connected to a definite high or low.

When using TTL logic, inputs are always pulled high with a resistor as opposed to being pulled low. If a resistor is used to pull an input low, as shown in Figure 47-10, it causes the gate to have a voltage drop at its output. This means that in the high state, the output of the gate may be only 3 or 4 volts instead of 5 volts. If this output is used as the input of another gate and this other gate has been pulled low with a resistor, the output of this second gate may be only 2 or 3 volts. Note that each time a gate is pulled low through a resistor, its output voltage becomes low. If this is done through several steps,

FIGURE 47-10 *Input pulled low with a resistor*

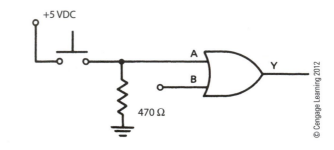

FIGURE 47-11 *Resistor used to pull the input of a gate high*

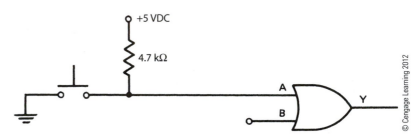

the output voltage soon becomes so low it cannot be used to drive the input of another gate high.

PULLING THE GATE HIGH: Figure 47-11 shows a resistor used to pull the input of a gate high. In this circuit, the push button is used to connect the input of the gate to ground or low.

If desired, the push button can produce a high at the input instead of a low. This is done with the addition of an INVERTER, as shown in Figure 47-12. In this circuit, a pull-up resistor is connected to the input of an INVERTER. Because the input is high, its output produces a low at the A input of the OR gate. When the normally open push button is pressed, it causes a low to be produced at the input of the INVERTER. When the input becomes low, its output becomes high. Note that the push button causes a high to appear at the A input of the OR gate when it is pushed.

Because both of the push buttons and the normally closed overload contact are used to provide high inputs, the circuit will be changed,

FIGURE 47-12 *Push button produces a high at the input.*

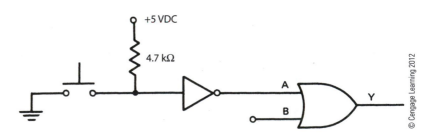

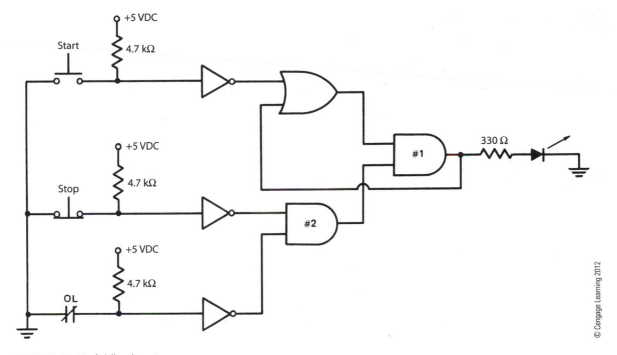

FIGURE 47-13 *Adding inverters*

as shown in Figure 47-13. Note that the normally closed push button and overload switch connected to the inputs of AND gate #2
are connected to ground instead of Vcc. When they are connected
to ground, a low is provided to the input of the INVERTERS they
are connected to. The INVERTERS therefore produce a high at the
input of the AND gate. If one of these normally closed switches is
opened, a high is provided at the input of the INVERTER gate. This
causes its output to become low. If the logic of the circuit shown in
Figure 47-13 is checked, it can be seen that it is the same as the logic
of the circuit shown in Figure 47-9

Correcting Faults: The Output

The final design problem for this circuit concerns the output. So far,
an LED has been used as the load. It is used to indicate when the
output is high and low. The original circuit, however, was used to
control a 120-volt AC motor.

UNIT 47 REVIEW QUESTIONS

1. In a relay circuit, what is used to maintain the operation of the circuit after the start button has been pushed? _____

2. What device is used by the overload relay to sense motor current? _____

3. In Figure 47-14, how is the circuit maintained after the start push button is released? _____
_____

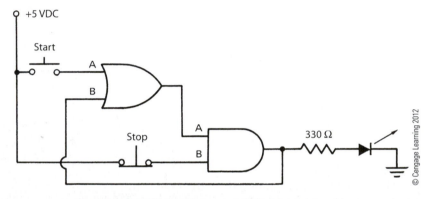

FIGURE 47-14 *Output turns on and off with push buttons*

4. Why should the input of a gate not be left disconnected? _____
_____
_____

5. Why are the inputs to gate taken high through a resistor instead of low through a resistor?
_____
_____

UNIT 48

DESIGN OF CIRCUIT #4
(Electronic Lock)

OBJECTIVES

After studying this unit, the student should be able to:

- Discuss the operation of an electronic lock.
- Construct an electronic lock using digital-logic gates.

CIRCUIT REQUIREMENTS

In this circuit, logic gates will be used to design an electronic lock. The requirements are as follows:

1. The opening of the lock occurs by pressing four numbers in the proper sequence.

2. All push buttons are normally open and one side of each button is connected to ground. This permits the use of a 10-digit keyboard with a common ground connection.

3. If any number that is not used in the combination is pressed, it will reset the circuit, making it necessary to begin again with the first number of the sequence.

For this circuit, the numbers 2, 4, 6, and 8 will be used as the combination.

NAND GATES USED AS A LATCH

The first part of this circuit can be constructed with two NAND gates used as a latch, Figure 48-1.

Understanding the Circuit

To understand the logic of this circuit, assume the output of NAND gate #1 to be low. This provides a low to the A input of NAND gate #2.

FIGURE 48-1 *Latch circuit*

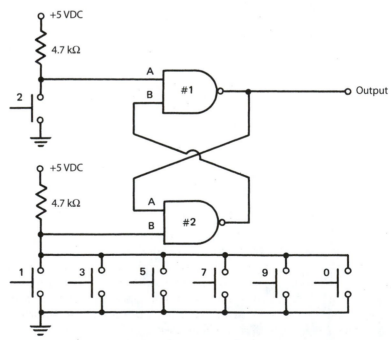

Because an NAND gate has a high output if any input is low, the output of NAND gate #2 provides a high for the B input of NAND gate #1. The A input of NAND gate #1 is forced high by the 4.7-kΩ resistor connected to +5-volts DC. The output of NAND gate #1 remains low and the output of NAND gate #2 remains high. If push button #2 is pressed, the A input of NAND gate #1 is driven low. This forces the output of NAND gate #1 to change to a high state. When the A input of NAND gate #2 changes to high, its output becomes low and provides a low to input B of NAND gate #1. When the B input of NAND gate #1 is forced low, the output of NAND gate #1 remains in a high state when push button #2 returns to its open position. When the output of NAND gate #1 changes to a high state, it forces the A input of the NAND gate #2 to also become high. NAND gate #2 now has both of its inputs high. This forces the output to remain low.

If push button #1, 3, 5, 7, 9, or 0 is pushed, the B input of NAND gate #2 is connected to ground or low. When the B input of NAND gate #2 becomes low, its output changes to a high state. This provides a high to the B input of NAND gate #1. Both inputs of NAND gate #1 are now high. This forces the output to change to a low state. When the output of NAND gate #1 becomes low, it provides a low to the A input of NAND gate #2. Because the A input of NAND gate #2 is now low, it forces the output to remain high when push button #1, 3, 5, 7, 9, or 0 is released and returns to its open position.

Fulfilling Requirement #3

If the output of NAND gate #1 is used as the circuit output, a circuit exists that has a high output when push button #2 is pressed, and a low output when push button #1, 3, 5, 7, 9, or 0 is pressed. This fulfills the part of the circuit requirement that states that if any push button not used in the combination is pressed, it will reset the circuit. It is necessary to begin again with the first number of the combination sequence. The circuit in Figure 48-1 shows the connection of the six push buttons not used in the combination sequence. For the remainder of this unit, one push button, labeled N, will be used to represent these six push buttons.

LATCHING WITH A SINGLE PUSH BUTTON

The next problem is to design a circuit that can be latched with a single push button only after it receives a high input from the output of NAND gate #1. This would normally be considered a simple circuit to design, but the requirement of using push buttons with common grounds makes it more difficult than it first appears. There are probably several ways this can be accomplished, but the circuit shown in Figure 48-2 is the circuit that will be used.

FIGURE 48-2 *Latching with a single push button*

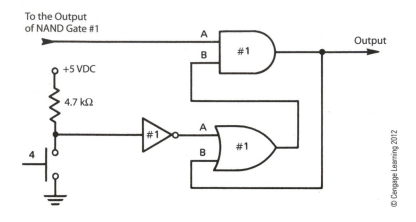

Circuit Logic

To understand the logic of this circuit, assume the output of the AND gate to be low. If the output of the AND gate is low, the B input of the OR gate is also low. The A input of the OR gate is connected to the output of an INVERTER. The input of the INVERTER is connected to high through a 4.7-kΩ resistor connected in turn to +5-VDC. Because the input of the INVERTER is high, its output is low. With both inputs of the OR gate low, its output provides a low to the B input of the AND gate. Because this input is low, the output is forced to remain low even if the A input is provided with a high from the output of NAND gate #1. Now assume that the A input of the AND gate has been driven high by the output of NAND gate #1. If push button #4 is pressed, the input of the INVERTER is connected to ground or low. This forces the output of the INVERTER to change and provides a high to the A input of the OR gate. When this input becomes high, the output of the OR gate changes and provides a high to the B input of the AND gate. Because both inputs of the AND gate are now high, its output changes from a low to a high. When this output becomes high, it provides a high to the B input of the OR gate. The high connected to the B input forces the output of the OR gate to remain high when push button #4 is released and the A input becomes low again. Once the AND gate has been turned on, its output remains in the high state until the output of NAND gate #1 changes state and forces the A input of the AND gate to change to a low state.

Creating the Combination

An electronic lock now exists with a two-number combination. If an output is to be obtained at the AND gate, push button #2 must first be pressed, followed by push button #4. Using the circuit shown in Figure 48-2 as a building block, it becomes a simple matter to add the other two numbers of the combination. When the output of the

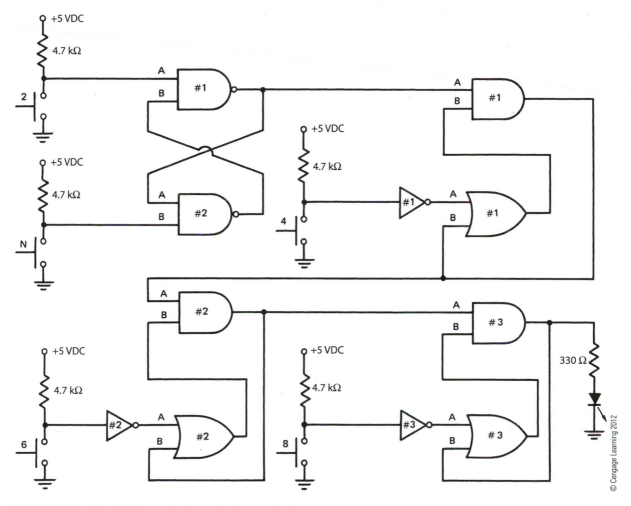

FIGURE 48-3 *Added latching circuits create the combination.*

AND gate in Figure 48-2 is used as the input of another identical circuit, the rest of the circuit becomes a simple chain. In the circuit shown in Figure 48-3, the input of a block is provided by the output of the block preceding it. For instance, AND gate #3 cannot have a high output unless AND gate #2 provides a high for its A input. AND gate #2 cannot have a high output unless AND gate #1 provides a high to its A input. Note that each of the push buttons used in the combination must be pressed in the proper sequence before AND gate #3 provides a high output.

This circuit uses an LED to indicate a high or low output. In actual practice, however, the output of AND gate #3 could be connected to the base of a transistor. The transistor could be used to control the coil of a relay, or the output of AND gate #3 could be connected to the input of a solid-state relay. The solid-state relay could then be used to control almost anything.

CIRCUIT CONNECTION

Until now the logic circuits that have been connected were relatively simple, using only two or three gates. If the INVERTERS are counted as gates, this circuit requires a connection of 11 different gates. Although only 4 ICs are required to construct the circuit, it is easy to become confused when making so many connections in a small space. It is helpful to number the schematic with the pin numbers of the IC. If a connection diagram of the IC is used, as shown in Unit 43, it is a simple matter to number the gate connections with the proper pin numbers of the IC, Figure 48-4.

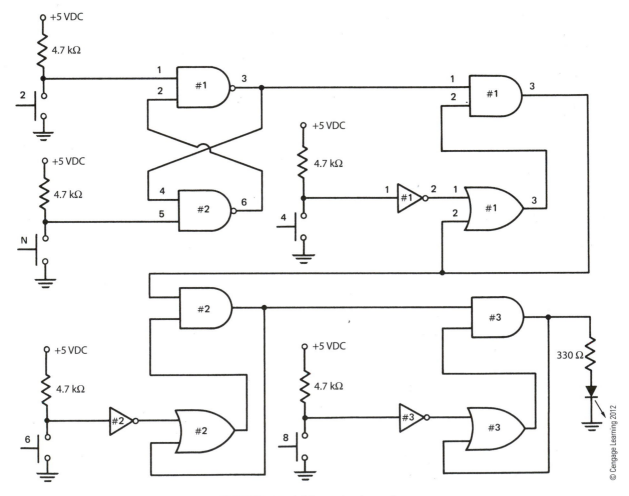

FIGURE 48-4 *Adding gate pin numbers*

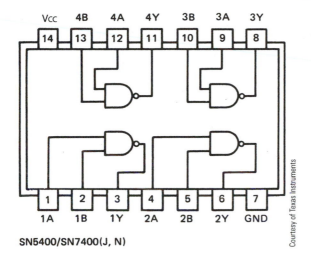

FIGURE 48-5 *NAND gate IC*

SN5400/SN7400(J, N)

Numbering the Gates

In Figure 48-5, the connection diagram for a 7400N IC is shown. The 7400N is a quad two-input NAND gate. Note in Figure 48-5 that NAND gate #1 has been labeled with a #1 and #2 at its inputs and a #3 at its output. The connection diagram in Figure 48-5 shows that one of the NAND gates in the 7400 IC has its inputs connected to pins #1 and #2 and its output is connected to pin #3. NAND gate #2 in Figure 48-4 has its inputs labeled #4 and #5 and its output is labeled #6. The connection diagram in Figure 48-5 shows that the second gate in the 7400 IC has its inputs connected to pins #4 and #5. Its output is connected to pin #6.

If each gate is labeled in this manner, connection becomes much easier. Figure 48-6 shows the connection diagrams for a 7408, quad two-input AND gate; a 7432, quad two-input OR gate; and a 7404, hex INVERTER. These connection diagrams have been used to label the input and output pins of AND gate #1, OR gate #1, and INVERTER #1. Figure 48-4 shows that pin #1 of the 7400 IC should be connected to the junction point of the 4.7-kΩ resistor and push button #2. Pin #3 should be connected to pin #4 of the 7400 IC and pin #1 of the 7408 IC. Pin #6 of the 7400 IC connects to pin #2 of the 7400 IC.

Note that connection becomes much simpler if the schematic is first labeled with the proper IC pin numbers. They are then used as a guide when connecting the circuit.

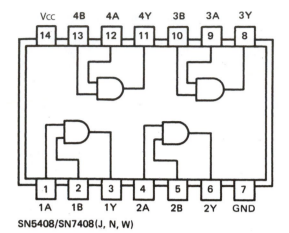

SN5408/SN7408(J, N, W)

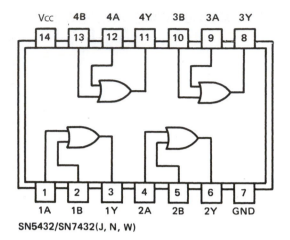

SN5432/SN7432(J, N, W)

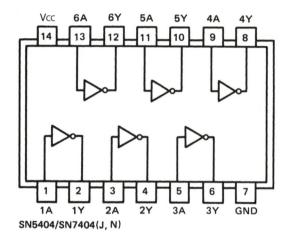

SN5404/SN7404(J, N)

FIGURE 48-6 *IC diagrams*

UNIT 48 REVIEW QUESTIONS

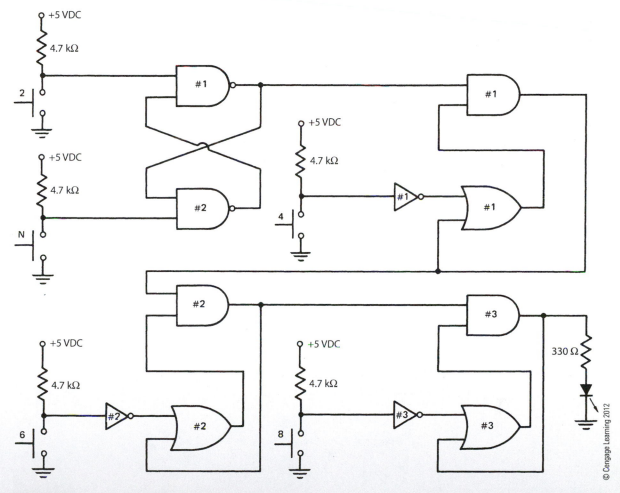

FIGURE 48-7

Refer to Figure 48-7 to answer the following questions.

1. What happens when push button N is pressed? _____

_____

_____

2. When push button #4 is pressed, does the input to OR gate #1 become high or low? _____

_____

UNIT 48 REVIEW QUESTIONS

3. What is the function of the LED connected to the output of AND gate #3? _____

_____

4. What is the function of the 330-Ω resistor connected in series with the LED? _____

_____

5. Refer to the connection of NAND gates #1 and #2. What circuit discussed in a previous unit uses this connection? _____

_____

UNIT 49

DESIGN OF CIRCUIT #5
(Solid-State Thermostat)

OBJECTIVES

After studying this unit, the student should be able to:

- Gain practical experience using a 741 operational amplifier.

- Use a PN junction as a temperature-sensing device.

- Use a solid-state relay to control an AC load.

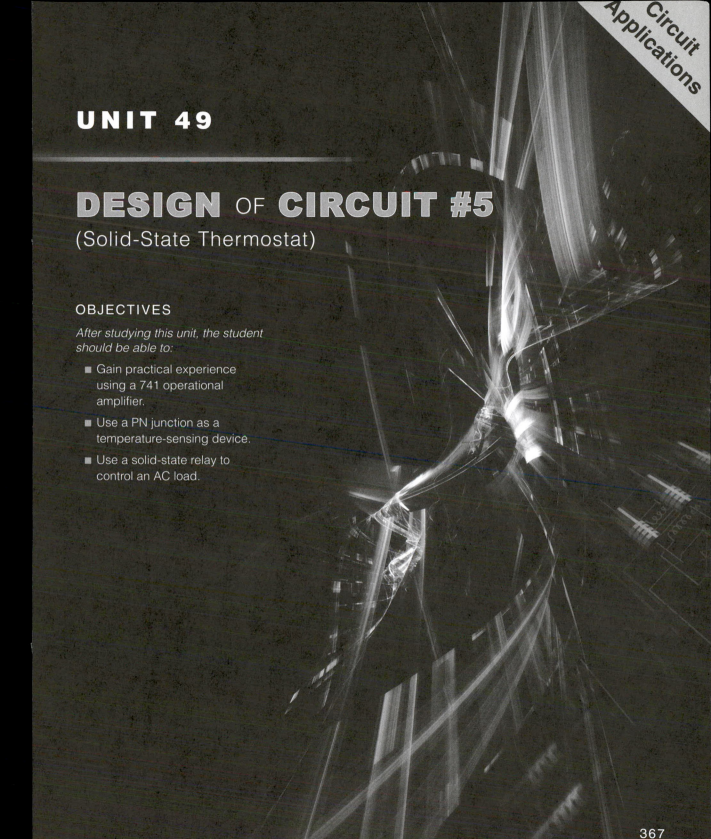

The last circuit to be designed is an electronic thermostat. Originally designed to control the temperature of a water bed, the heat sensor was placed on contact with the mattress and its output controlled the operation of an electric heater located under the mattress. Junction diodes are used as heat sensors, and the output of a 741 operational amplifier controls the input of a solid-state relay, which in turn controls the electric heater. The circuit is sensitive enough to maintain the temperature within 1° F.

JUNCTION DIODES USED AS HEAT SENSORS

Recall from previous units that semiconductor materials exhibit a change of resistance with a change of temperature and have a negative temperature coefficient, which means that their resistance will decrease with an increase of temperature. Junction diodes can be used as temperature sensors by passing a constant current through them and measuring the voltage drop across the PN junction. If the current flow through a diode is held constant, the voltage drop across the device will be proportional to its temperature. The voltage drop of a diode is not only proportional to temperature, the change is linear. Because semiconductor materials have a negative temperature coefficient, an increase of temperature will produce a decrease in voltage across the diode. Conversely, a decrease of temperature will produce an increase in the voltage drop.

In order for the diode to produce a voltage drop that is proportional to temperature, it must have a constant current passed through it. Therefore, a constant-current generator circuit is used to provide the current needed, Figure 49-1. In this circuit, a junction field effect transistor will be used to produce the constant current to the junction diode.

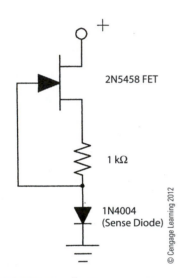

2N5458 FET

1 kΩ

1N4004 (Sense Diode)

© Cengage Learning 2012

FIGURE 49-1 *Constant-current generator supplies current to the sense diode.*

CIRCUIT CONSTRUCTION

The first step in developing this circuit will be to provide the power needed to operate the circuit. Because this circuit will use an operational amplifier as the main control device, an above- and below-ground power supply will be used. The power-supply circuit is shown in Figure 49-2. A 12.6 VCT (volts center tapped) transformer is used to step the 120 VAC down to 12.6 VAC. The center tap of the transformer is grounded, and a bridge rectifier is connected across the secondary windings of the transformer. The bridge rectifier will produce a voltage that is both negative and positive with respect to the center tap (ground). Two 1000-μF capacitors will be used to filter the DC. This power supply will produce approximately +9 and −9 VDC with respect to ground (6.3 V × 1.414 = 8.9 V).

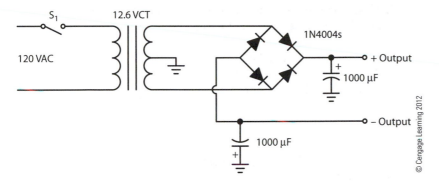

FIGURE 49-2 *Above- and belowground power supply*

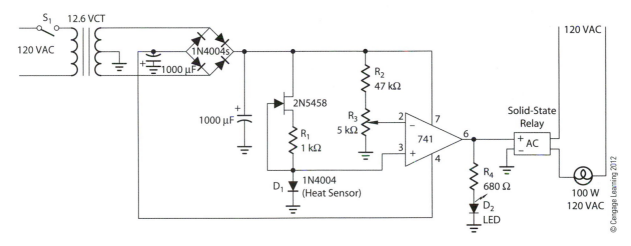

FIGURE 49-3 *Solid-state thermostat*

The outputs of the power supply are connected to the inputs of the op amp. The positive voltage source will be connected to pin #7 and the negative source will be connected to pin #4, Figure 49-3. The output of the 741 is connected to a solid-state relay that controls the 120-VAC heater. In this example, a 100-watt, 120-VAC lamp will be used to simulate the heater. A 680 Ω current-limiting resistor and light-emitting diode are also connected to the output of the op amp. The LED is used as an indicator to show when the output is turned on or off. In this circuit, the 100-watt lamp indicates when the output is on or off, but if the solid-state relay is used to control a resistive heating element, as the circuit is intended, a means is needed to indicate the state of the output.

CONNECTING THE INVERTING AND NONINVERTING INPUTS

This circuit is intended to be used as a heating thermostat. Therefore, the operational amplifier must be connected in such a manner that a drop in temperature will produce a positive voltage at the output. Because the resistance of semiconductor material is inversely proportional to temperature, the voltage drop across the device will decrease with an increase of temperature. The noninverting input will therefore be connected to the heat-sensing resistor.

The inverting input is connected to a voltage-divider circuit that is used to permit a greater amount of control at the inverting input. The voltage divider is comprised of resistors R_1 and R_3. Resistor R_3 is a variable resistor with a value of 5kΩ and a total voltage drop of approximately 0.86 volt. The full range of the variable resistor is used to control a voltage of 0–0.86 volt at the inverting input instead of 0–9 volts. Because the voltage drop of the heat-sensing diode will be between 0.3 volt and 0.7 volt, the voltage-divider circuit permits a much greater degree of control with the variable resistor.

CIRCUIT OPERATION

To understand the operation of this circuit, assume that the temperature of the heat-sensing resistor is lower than desired and the output of the op amp is off or low. The variable resistor is turned in such a manner so as to reduce the voltage at the inverting input. When the voltage at the inverting input becomes less than the voltage of the noninverting input, the output will turn on. This causes the heating element to heat the water in the mattress. As the water temperature increases, the voltage drop across the silicon diode decreases. When the voltage at the noninverting input becomes less than the voltage of the inverting input, the output of the op amp will go low or turn off.

As the water cools, the voltage drop across the sense diode will begin to increase. When the voltage drop becomes greater than the voltage applied to the inverting input, the output will again turn on and permit the heating element to heat the water. The water temperature can be controlled by adjusting the variable resistor.

Making a Cooling Thermostat

The thermostat circuit shown in Figure 49-3 can be changed from a heating to a cooling thermostat by reversing the connection of the inverting and noninverting inputs, Figure 49-4. This will permit the output of the op amp to turn on with an increase of temperature and

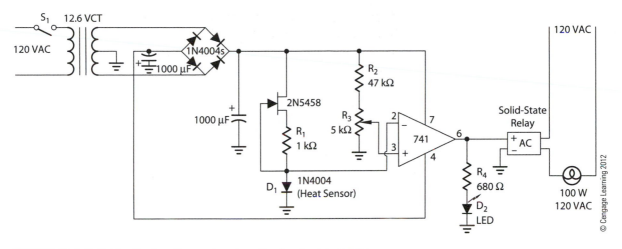

FIGURE 49-4 *Switching the inverting and noninverting inputs changes the circuit from a heating to a cooling thermostat.*

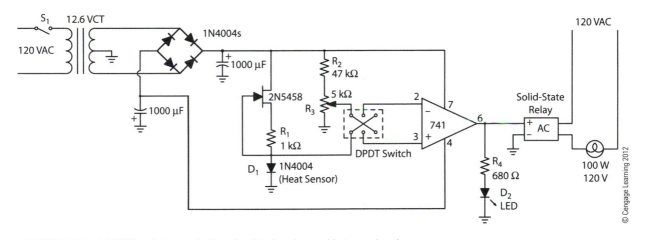

FIGURE 49-5 *A DPDT switch permits the circuit to be changed between heating and cooling modes.*

turn off with a decrease of temperature. If a double-pole, double-throw switch with crisscross connection is added to the circuit, the thermostat can be switched between heating and cooling modes, Figure 49-5.

UNIT 49 REVIEW QUESTIONS

1. Explain how to use a junction diode in such a manner that it can be used to sense a change of temperature. _____

_____

_____

2. Does an increase of temperature cause the voltage drop across a PN junction to increase or decrease?

_____

_____

3. Is the voltage drop across a junction diode linear or nonlinear with a change of temperature?

_____

_____

4. What must be done to change the thermostat from a heating mode to a cooling mode? _____

_____

5. What device was added to the circuit in Figure 49-5 to permit the thermostat to be changed between heating and cooling modes? _____

_____

LABORATORY
EXERCISES

FOREWORD

This textbook approaches the subject of electronics from a practical standpoint instead of a mathematical base. These experiments are intended to give students hands-on experience with many of the more common electronic components found in industry as opposed to components generally used in other electronic applications. The experiments start at a low level and progress to more difficult circuits. Many of these experiments are powered by a source of direct current (DC), which means that polarity must be observed when connecting certain components. This is especially true when connecting electrolytic capacitors. Electrolytic capacitors are very polarity sensitive and can explode if connected improperly. These capacitors will have some means of identifying the positive and negative leads. Some have an arrow pointing to the negative lead as shown in Figure F-1. The circles with the negative symbols represent negative electrons. Other electrolytic capacitors may have a plus sign close to one lead as shown in Figure F-2. Regardless of the method of identification, it is imperative that they be connected to the correct polarity. Capacitors that do not have polarity markings are generally not polarity sensitive and are often referred to as AC capacitors. The leads of AC capacitors can be connected to either polarity without harm.

Another area of concern for many students that are unfamiliar with electronic diagrams is the use of ground symbols, Figure F-3. Ground symbols are basically a form of short hand. The circuit shown in Figure F-4 contains many points of ground. All grounds are to be connected together. When all grounds are connected, it is the same as the circuit shown in Figure F-5. Large electronic diagrams may contain several hundred ground symbols. It is much simpler to draw a ground symbol than to show connection of all the components.

FIGURE F-1 *Electrolytic capacitors have some means of identifying the positive and negative leads.*

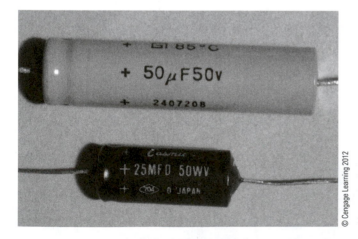

FIGURE F-2 *The positive lead is identified with a plus sign.*

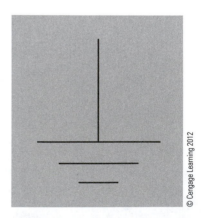

FIGURE F-3 *Ground symbol used in many electronic circuits.*

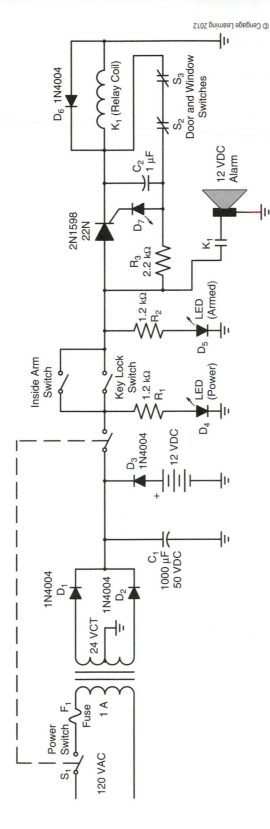

FIGURE F-4 *Ground symbols are used as a type of short hand.*

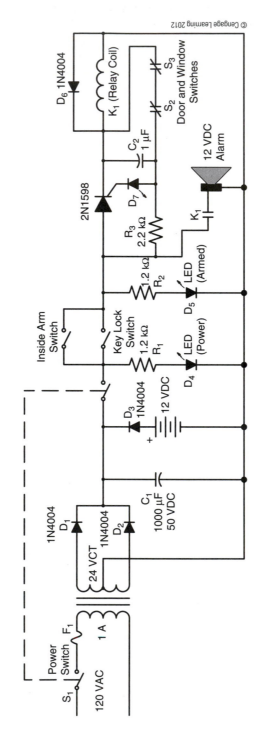

FIGURE F-5 *All ground points are connected together.*

Resistors are one of the most common components found in the electronic field. The resistors used in these experiments are rated at ½ watt unless otherwise noted. Resistor with a higher wattage rating than that listed may be used without problem, but resistors with a lower wattage rating will probably be damaged. Some electronic components have leads that are larger than others. A 2 watt resistor, for example, has leads that are larger than a ½ watt resistor. The larger size leads often will not fit into the breadboards used for low power applications. A company called Energy Concepts Inc. does make a bread boarding system that can be used with components with larger leads.

Many of the experiments contain an isolation transformer. These transformers have a turns ratio of 1:1 and produce the same voltage at the secondary as the primary. The purpose of this transformer is to remove the circuit from ground. Oscilloscopes have a grounding lead that is attached to the case of the scope. If the ground lead is mistakenly connected to the ungrounded side of the circuit, it will produce a short and generally destroy the oscilloscope lead. The isolation transformer prevents this possibility.

LAB EXERCISE 1 OSCILLOSCOPE

© Shutterstock 2012

_____ Volts   **1.** Connect the circuit shown in Exp 1-1.

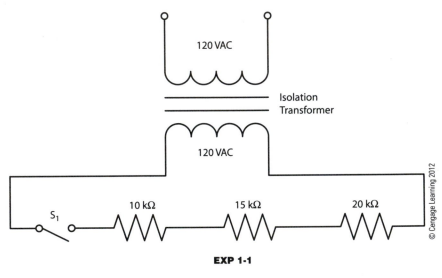

EXP 1-1

© Cengage Learning 2012

2. Turn on the oscilloscope and wait until a single line appears across the screen.

3. Switch the AC–GND–DC switch to the GND position.

4. Adjust the intensity. Focus controls to produce a sharp line and a medium brightness.

5. Set the volts-per-division adjustment for 20 volts. (Note: If using a 10:1 attenuator, set the volts-per-division switch for 2 volts.)

6. Adjust the vertical position control so the trace line is on the centerline of the X-axis.

7. Set the time base for a value of 5 milliseconds per division. (Note: Be sure the trigger is set for auto and the scope is set to trigger on the channel being used. For example, if the scope probe is connected to channel #1, be sure the trigger is set for channel #1.) Set the scope to trigger on the positive slope of the waveform.

8. Set the AC–GND–DC switch for DC.

9. Connect the oscilloscope probe across the 10-kΩ (kilohm) resistor.

10. Turn on switch S_1 and observe the voltage on the display of the oscilloscope. Measure the peak voltage of the waveform shown on the display by counting the number of lines up from the centerline of the display. Multiply this number by 20 because the scope is set for 20 volts per division. (Note: The horizontal position control can be used to move the trace so that the peak of the waveform is on the center vertical line. This will permit the use of hash marks, shown on the center vertical line, to obtain a more accurate measurement of the peak of the waveform when it is between the major divisions shown on the display.)

_____ Volts

11. The voltage measured in step #10 is the peak of an AC sine wave. To find the rms value, multiply the voltage measured by 0.707.

_____ Volts

12. Measure the voltage across the 10-kΩ resistor with an AC voltmeter.

_____ Volts

13. Compare the value of voltage in step #11 with the voltage measured in step #12.

14. To measure the frequency of the waveform, adjust the horizontal position control until the point at which the waveform is beginning to rise in the positive direction is positioned to the first vertical line at the left of the oscilloscope display. Figure 1-5 shows a waveform in this position.

15. Measure the time it takes the waveform to complete one cycle by counting the divisions on the X-axis it takes the waveform to complete one cycle and multiply this by 0.005 seconds per division. Divide this answer into 1 ($F = 1/t$).

Frequency _____

16. Change the setting of the time base to 2 milliseconds per division.

17. Compute the frequency of the waveform by measuring the time it takes to complete one cycle and divide this into 1 ($F = 1/t$).

Frequency _____

18. Open switch S_1 and reconnect the oscilloscope probe across the 15-kΩ resistor.

19. Close switch S_1 and measure the peak voltage dropped across this resistor.

_____ Volts

20. Compute the rms voltage by multiplying the voltage in step #19 by 0.707.

_____ Volts

21. Measure the voltage dropped across the 15-kΩ resistor with an AC voltmeter.

_____ Volts

22. Compare the value of voltage computed in step #20 with the voltage measured in step #21.

23. Measure the frequency of the waveform shown on the display.

Frequency _____

24. Open switch S$_1$ and reconnect the oscilloscope probe across the 20-kΩ resistor.

_____ Volts

25. Close switch S$_1$ and measure the peak voltage shown on the display.

_____ Volts

26. Compute the rms voltage by multiplying the voltage in step #25 by 0.707.

_____ Volts

27. Measure the voltage dropped across the 20-kΩ resistor with an AC voltmeter.

28. Compare the voltage computed in step #26 with the voltage measured in step #27.

Frequency _____

29. Measure the frequency of the waveform shown on the display.

LAB EXERCISE 2   JUNCTION DIODES

1. Test several diodes with an ohmmeter in the manner described in this unit.

2. Connect the circuit shown in Exp 2-1.

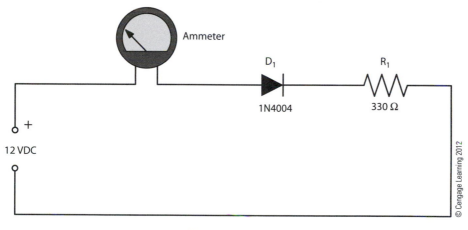

EXP 2-1

3. Using a digital voltmeter, measure the voltage drop across diode D_1.   _____ Volts

4. Measure the current flow in the circuit.   _____ A

5. Replace resistor R_1 with a 2.2-kΩ resistor.

6. Measure the voltage drop across diode D_1.   _____ Volts

7. Measure the current flow in the circuit.   _____ A

8. Compare the voltage drop in step #3 with the voltage drop in step #6.

9. Reverse the direction of diode D_1.

10. Measure the voltage drop of diode D_1.

11. Measure the current flow in the circuit.   _____ A

LAB EXERCISE 3   LIGHT-EMITTING DIODES

1. Connect the circuit shown in Exp 3-1.

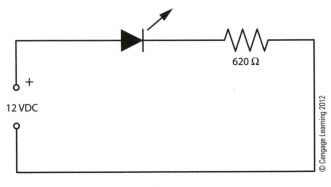

620 Ω

+

12 VDC

EXP 3-1

_____ Volts   2. Connect a digital voltmeter across the LED. Record the voltage drop of the device.

3. Reverse the connection of the LED. Observe whether the LED lights.

LAB EXERCISE 4   SINGLE-PHASE RECTIFIERS

1. Connect the circuit shown in Exp 4-1.

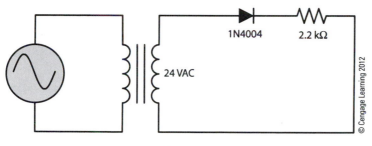

EXP 4-1

2. Measure the AC voltage at the secondary of the transformer. _____ Volts

3. Compute what the DC voltage across the load resistor should be: VAC × 0.9/2 = VDC. _____ Volts

4. Measure the voltage across the load resistor. Compare it to your computed value. _____ Volts

5. Connect an oscilloscope across the load resistor. Draw the waveform that appears on the display.

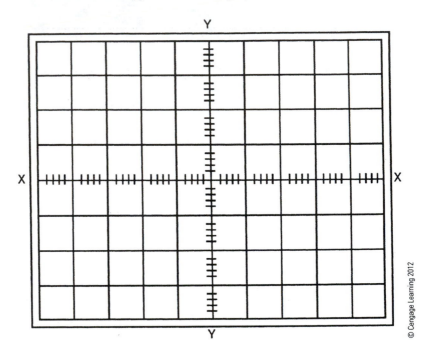

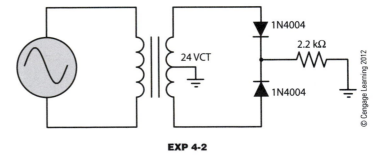

EXP 4-2

6. Connect the circuit shown in Exp 4-2.

_____ Volts

7. Measure the AC voltage of the secondary of the transform between the center tap lead and one of the other secondary leads.

_____ Volts

8. Compute the output voltage of the rectifier (VAC × 0.9 = VDC).

_____ Volts

9. Measure the voltage produced across the load resistor. Compare it to your computed value.

10. Connect an oscilloscope across the load resistor. Draw the waveform on the display.

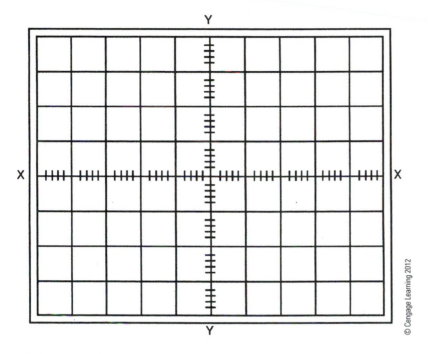

© Cengage Learning 2012

11. Connect the circuit shown in Exp 4-3.

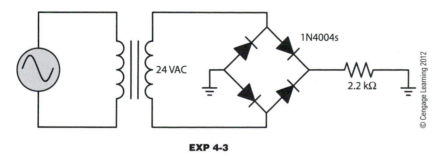

1N4004s

24 VAC

2.2 kΩ

© Cengage Learning 2012

EXP 4-3

12. Measure the AC voltage of the transformer secondary.   _____ Volts

13. Compute the average DC voltage output of the rectifier   _____ Volts
(VAC × 0.9 = VDC).

14. Measure the voltage across the load resistor. Compare the   _____ Volts
measured voltage with the computed value.

15. Connect an oscilloscope across the load resistor. Draw the waveform shown on the display.

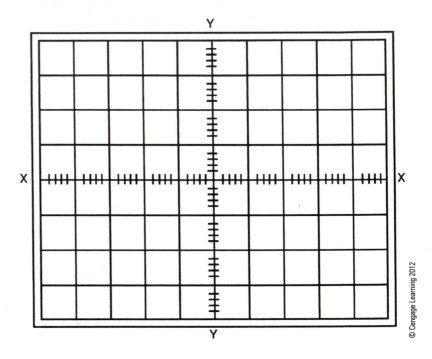

LAB EXERCISE 5

THREE-PHASE RECTIFIERS

▶ CAUTION: High voltages will be used in this experiment.

1. Connect the circuit shown in Exp 5-1.

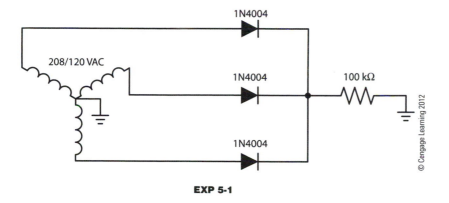

EXP 5-1

2. Connect a DC voltmeter across the load resistor.

3. Turn on the three-phase power. Record the voltage across the load resistor.   _____ Volts

4. Turn off the AC power. Connect an oscilloscope across the load resistor.

5. Turn on the AC power. Draw a picture of the AC waveform shown on the oscilloscope.

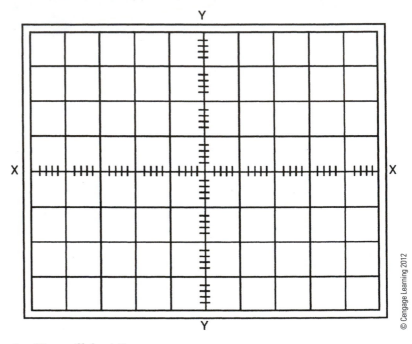

6. Turn off the AC power.

7. Connect the circuit shown in Exp 5-2.

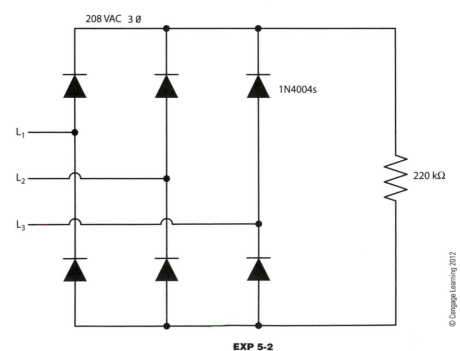

EXP 5-2

8. Connect a DC voltmeter across the load resistor.

_____ Volts

9. Turn on the AC power. Record the voltage indicated by the voltmeter.

10. Turn off the AC power. Connect an oscilloscope across the load resistor.

11. Turn on the AC power. Draw a picture of the waveform seen on the display of the oscilloscope.

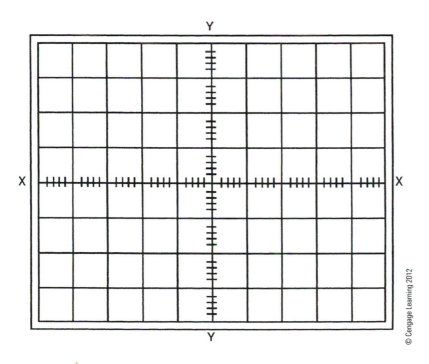

© Cengage Learning 2012

12. Turn off the AC power.

LAB EXERCISE 6   FILTERS

1. Connect the circuit shown in Exp 6-1.

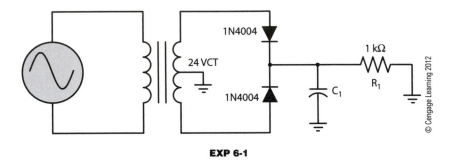

EXP 6-1

2. Connect a voltmeter and an oscilloscope across R_1.

3. By changing the value of C_1, fill in the chart. Each time the capacitor is changed, observe the change of the waveform shown on the oscilloscope.

C_1	VOLTS
1 μF	
25 μF	
100 μF	
150 μF	
200 μF	

4. Replace resistor R_1 with a 100-Ω, 5-watt resistor.

5. By changing the value of capacitance, fill in the chart.

C_1	VOLTS
10 μF	
50 μF	
100 μF	
500 μF	
1000 μF	

LAB EXERCISE 7   ZENER DIODES

1. Test the zener diode with an ohmmeter to insure that it is good.

2. Connect the circuit shown in Exp 7-1.

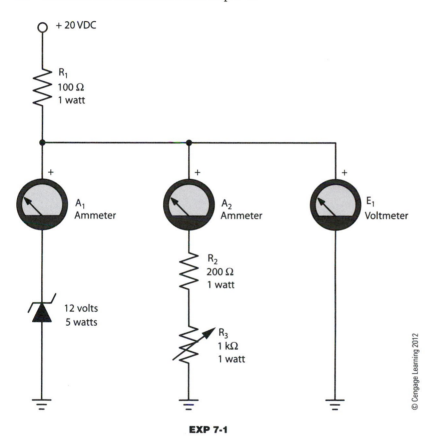

+ 20 VDC

R_1
100 Ω
1 watt

A_1
Ammeter

A_2
Ammeter

E_1
Voltmeter

R_2
200 Ω
1 watt

12 volts
5 watts

R_3
1 kΩ
1 watt

EXP 7-1

3. Fill in the unknown values in the chart by adjusting the value of R_3 to obtain the currents shown for meter A_2.

E_1	A_1	A_2
		0.05
		0.04
		0.03
		0.02

LAB EXERCISE 8

TRANSISTORS

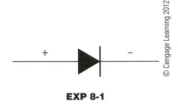

EXP 8-1

1. Using an ohmmeter, determine which lead of the meter is positive and which is negative. This can be done by connecting the ohmmeter leads to a junction diode. The diode shows continuity only when the leads of the ohmmeter are connected, as shown in Exp 8-1.

2. Test several transistors with the ohmmeter. Determine whether they are good or bad and whether they are NPN or PNP.

3. Using a plastic case transistor, find the type of transistor. Identify each lead as being the base, collector, or emitter.

4. Connect the circuit shown in Exp 8-2.

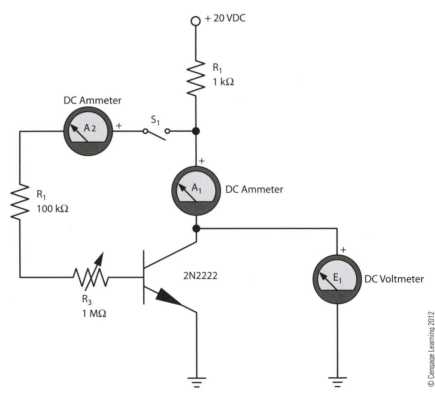

EXP 8-2

5. Fill in the chart by adjusting the resistor R_3 to obtain the proper base current shown in the chart. Find the impedance by dividing the voltage drop across the transistor by the current of meter A_1.

BASE CURRENT (A_2)	COLLECTOR CURRENT (A_1)	VOLTAGE DROP	IMPEDANCE
0 (SWITCH OPEN)			
20 μA			
40 μA			
60 μA			
80 μA			

394 ELECTRONICS FOR ELECTRICIANS

LAB EXERCISE 9   TRANSISTOR SWITCH

© Shutterstock 2012

1. Connect the circuit shown in Exp 9-1.

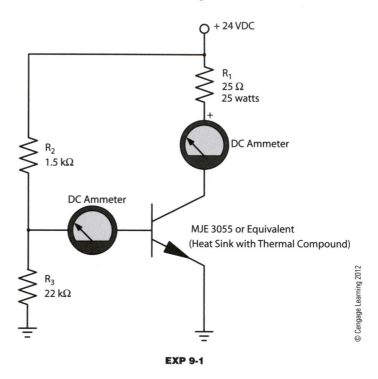

EXP 9-1

© Cengage Learning 2012

2. Note the difference in current flowing through the base and the collector–emitter of the transistor.

_____ Volts

3. Measure the voltage drop across the collector and the emitter of the transistor with a digital voltmeter.

(Note: Resistor R_1 is considered the load. It limits the current in the collector–emitter section of the transistor. Resistor R_2 determines the amount of current permitted to flow to the base of the transistor. Resistor R_3 is used to insure that the transistor will turn off when the base is disconnected from the positive supply voltage. R_3 is necessary because transistors are sometimes sensitive enough to be triggered by magnetic interference in the air.)

◆ CAUTION: The transistor must be mounted on a heat sink and thermal compound used.

LAB EXERCISE 10   TRANSISTOR AMPLIFIER

1. Connect the circuit shown in Exp 10-1.

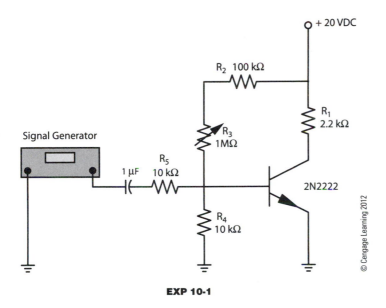

EXP 10-1

2. With the signal generator disconnected, adjust resistor R_3 until there is a 10-volt drop across the transistor.

3. Connect a signal generator to the circuit. Connect the oscilloscope across the collector–emitter of the transistor. Adjust the signal generator until a sine-wave voltage appears on the oscilloscope display.

4. Measure the peak-to-peak voltage of the waveform appearing on the oscilloscope.   _____Vpp

5. Connect the oscilloscope across the base–emitter of the transistor. Measure the peak-to-peak voltage.   _____Vpp

6. Compute the amplification of the circuit by dividing the peak-to-peak voltage dropped across the transistor by the peak-to-peak voltage applied to the base of the transistor.   _____ Gain

LAB EXERCISE 11   DARLINGTON AMPLIFIER

1. Connect the circuit shown in Exp 11-1.

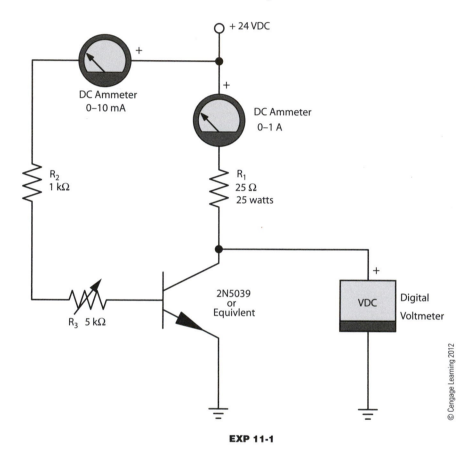

EXP 11-1

2. Decrease the resistance of resistor R_3 until the transistor is saturated and the voltage drop across the collector–emitter is about 0.3 volts.

A_1_____

3. Record the currents indicated by meters A_1 and A_2.

A_2_____

A_1/A_2 Gain_____

4. Compute the circuit gain by dividing the reading of meter A_1 by A_2.

5. Connect the circuit shown in Exp 11-2.

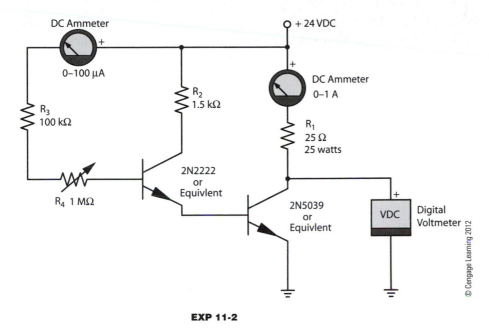

EXP 11-2

6. Decrease the resistance of resistor R_4 until transistor Q_1 is saturated. (Note: You can tell the transistor is saturated when the voltage drop across the collector and emitter is about 0.3 volts.)

7. Measure the current indicated by meters A_1 and A_2.

A_1 _____

A_2 _____

8. Compute the circuit gain by dividing the reading of meter A_1 by A_2.

A_1/A_2 Gain _____

9. Compare the gain of the circuit in exercise 1 to the gain of the circuit in exercise 5.

LAB EXERCISE 12

FIELD EFFECT TRANSISTORS

1. Connect the circuit shown in Exp 12-1.

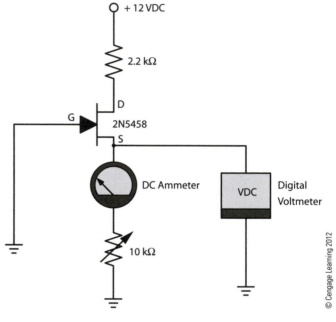

EXP 12-1

_____ Volts

2. Adjust the variable resistor until the maximum amount of current flows through the circuit. What is the amount of voltage between the source and gate?

_____ Volts

3. Adjust the variable resistor until there is no current flow in the circuit. What is the amount of voltage between the source and gate?

_____ Volts

4. Adjust the variable resistor until there is about one-quarter of the maximum current flowing in the circuit. What is the amount of voltage between the source and gate?

_____ Volts

5. Adjust the variable resistor until there is about one-half of the maximum current flowing in the circuit. What is the amount of voltage between the source and gate?

_____ Volts

6. Adjust the variable resistor until there is about three-quarters of the maximum current flowing the in circuit. What is the amount of voltage between the source and gate?

LAB EXERCISE 13   CURRENT GENERATORS

© Shutterstock 2012

1. Connect the circuit shown in Exp 13-1. The value of the load resistor is to be changed during the experiment. The values of resistance shown in the chart in exercise 4 are to be used for the load resistor.

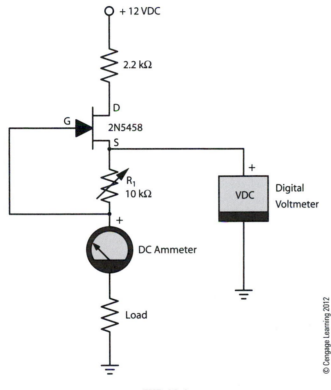

EXP 13-1

© Cengage Learning 2012

2. Turn on the power supply and adjust the voltage for a value of 12 volts.

3. Adjust resistor R_1 until there is a current of 1 mA flowing in the circuit.

4. Fill in the chart by measuring the current flow in the circuit and the voltage drop across the load resistor for each resistor shown in the chart.

RESISTANCE	100 Ω	1000 Ω	2200 Ω	3000 Ω
CURRENT				
VOLTS				

5. Adjust resistor R_1 until a current of 2 mA flows in this circuit. Fill in the chart by substituting the values of resistance listed for the load resistor.

RESISTANCE	100 Ω	1000 Ω	2200 Ω	3000 Ω
CURRENT				
VOLTS				

6. Turn off the power supply and reconnect the circuit as shown in Exp 13-2. The values of the load resistor will be changed during the experiment by substituting the values of resistance listed in the chart in exercise 8.

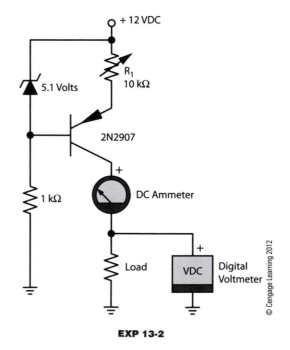

EXP 13-2

© Cengage Learning 2012

7. Turn on the power supply and adjust resistor R_1 for a current flow of 1 mA.

8. Fill in the chart in the figure by replacing the load resistor with the value of resistance listed in the chart.

RESISTANCE	100 Ω	1000 Ω	2200 Ω	3000 Ω
CURRENT				
VOLTS				

9. Adjust resistor R_1 until a current of 2 mA flows in the circuit.

10. Fill in the chart by substituting the values of resistance listed for the load resistor.

RESISTANCE	100 Ω	1000 Ω	2200 Ω	3000 Ω
CURRENT				
VOLTS				

11. Turn off the power supply and disconnect the circuit.

1. Connect the circuit shown in Exp 14-1.

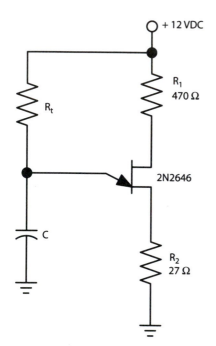

R_t	1 µF	0.1 µF	0.01 µF
10 kΩ	Hz	Hz	Hz
22 kΩ	Hz	Hz	Hz
47 kΩ	Hz	Hz	Hz

EXP 14-1

2. Resistor R_t is to be changed to the values indicated in the table.

3. Fill in the missing values in the table by substituting the values for R_t and C.

4. Connect an oscilloscope across resistor R_2. Measure the frequency of the output pulses. (Note: To measure frequency, measure the time between pulses and divide this time into 1.) Example: One pulse measures 50 microseconds.

$$F = \frac{1}{0.000050} = 20{,}000 \text{ Hz}$$

© Cengage Learning 2012

LAB EXERCISE 15   SCR CHARACTERISTICS

1. Connect the circuit shown in Exp 15-1.

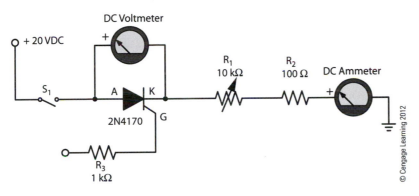

EXP 15-1

2. Set resistor R_1 to have a value of 0 Ω.

3. Close switch S_1. Record the voltage drop across the SCR and the current flow through the circuit.

E _____

I _____

4. Turn the SCR on by momentarily touching the open end of R_3 to the anode of the SCR.

5. Measure the voltage drop across the SCR and the current flowing in the circuit.

E _____

I _____

6. Slowly increase the resistance of R_1 until the SCR turns off. Measure the amount of current at this point to find the holding current value of the SCR.

I _____

LAB EXERCISE 16 SCR IN AN AC CIRCUIT

1. Connect the circuit shown in Exp 16-1.

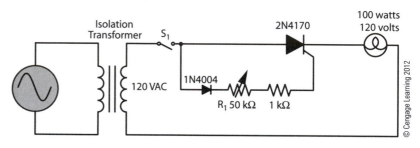

EXP 16-1

2. Set R_1 for maximum resistance.

3. Close switch S_1.

_____ Volts   4. Connect a DC voltmeter across the load. Slowly adjust R_1 until the SCR fires and the bulb initially lights. Record the voltage at this point.

_____ Volts   5. Slowly adjust resistor R_1. Watch the voltmeter while adjusting the resistor. When R_1 is adjusted to 0 Ω, record the voltage.

6. Open switch S_1.

7. If an oscilloscope is available, connect it across the bulb. Repeat the above procedure, watching the waveform shown on the oscilloscope.

8. Reset R_1 for its maximum value and open switch S_1.

9. Reconnect the oscilloscope across the SCR and slowly adjust R_1 for its zero value. Observe the point at which the SCR fires. Observe what happens as R_1 is adjusted toward its zero value.

LAB EXERCISE 17   PHASE SHIFTING AN SCR

1. Connect the circuit shown in Exp 17-1.

   (Note: The gate voltage is polarity sensitive. If the circuit does not operate correctly, change the polarity of the gate voltage by reversing the connection of the secondary leads of the transformer.)

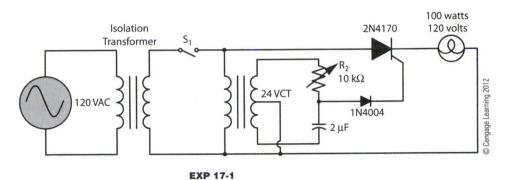

EXP 17-1

2. Adjust R_2 for its maximum value.

3. Close switch S_1.

4. Slowly adjust R_2 from its maximum value to its minimum value while watching the voltmeter connected across the load.

5. Return R_2 to its maximum value while observing the voltmeter.

6. Open switch S_1.

7. Connect an oscilloscope, if available, across the SCR. Repeat steps #3 through #6.

8. Reconnect the oscilloscope across the load. Repeat steps #3 through #6.

LAB EXERCISE 18

UJT PHASE SHIFTING OF AN SCR

1. Connect the circuit shown in Exp 18-1.

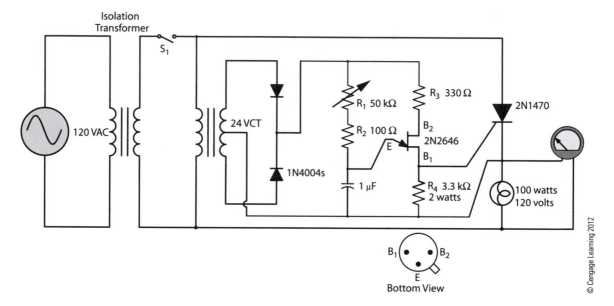

EXP 18-1

2. Set R_1 for its maximum value.

3. Close switch S_1.

4. Slowly decrease the resistance of R_1 to 0 Ω and then back to maximum while observing the voltmeter connected across the load.

5. Open switch S_1.

6. Connect an oscilloscope across the load. Repeat steps #3 through #5.

7. Connect the oscilloscope across the anode–cathode of the SCR. Repeat steps #3 through #5.

LAB EXERCISE 19

SCR FULL-WAVE RECTIFIER

© Shutterstock 2012

1. Connect the circuit shown in Exp 19-1.

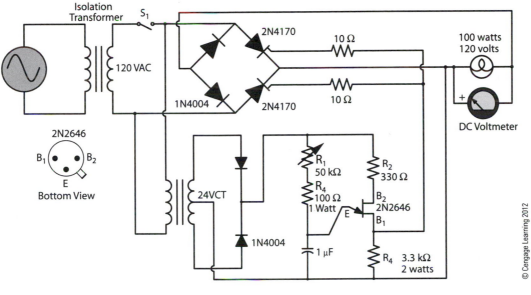

EXP 19-1

© Cengage Learning 2012

2. Set R_1 for its maximum value.

3. Close switch S_1.

4. Slowly adjust R_1 for a value of 0 Ω and back again to maximum resistance while observing the voltmeter connected across the load.

5. Open switch S_1.

6. Connect an oscilloscope across the load resistor. Repeat steps #3 through #5.

LAB EXERCISE 20   ALARM SYSTEM WITH BATTERY BACKUP

1. Connect the circuit shown in Exp 20-1.

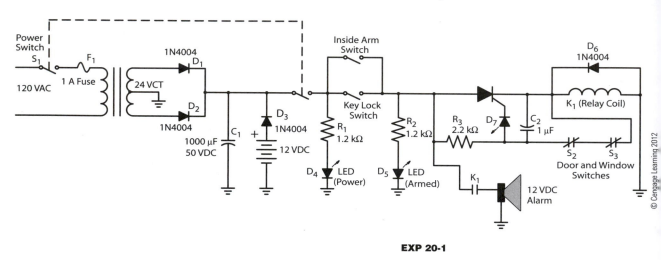

EXP 20-1

2. Test the operation of the circuit as described in this lesson.

LAB EXERCISE 21   DIAC CHARACTERISTICS

© Shutterstock 2012

1. Connect the circuit shown in Exp 21-1.

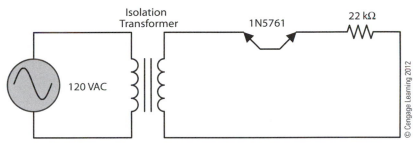

EXP 21-1

© Cengage Learning 2012

2. Connect an oscilloscope across the diac and observe the waveform.

3. Connect an AC voltmeter across the diac and record the voltage drop.

_____ Voltage drop

4. If possible, lower the AC voltage with an autotransformer. Observe the voltage drop across the diac.

(Note: Because the voltage drop across the diac remains constant regardless of the applied voltage, it can be used as an AC zener diode. Refer to Unit 9.)

LAB EXERCISE 22   TRIAC CHARACTERISTICS

1. Using an ohmmeter, test a triac in the manner described in this unit.

2. Connect the circuit shown in Exp 22-1.

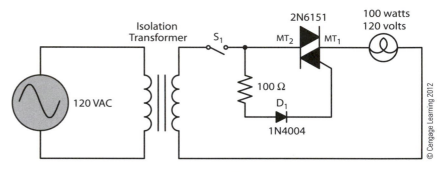

EXP 22-1

3. Connect an oscilloscope across the light bulb that is being used as the load resistor.

4. Close switch S_1.

5. Draw a picture of the waveform that appears on the screen of the oscilloscope.

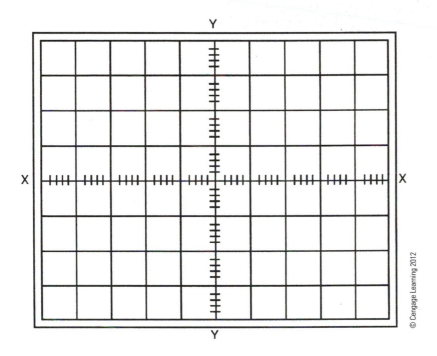

© Cengage Learning 2012

6. Open switch S_1.

7. Reverse the direction of diode D_1.

8. Close switch S_1.

9. Draw a picture of the waveform that appears on the oscilloscope.

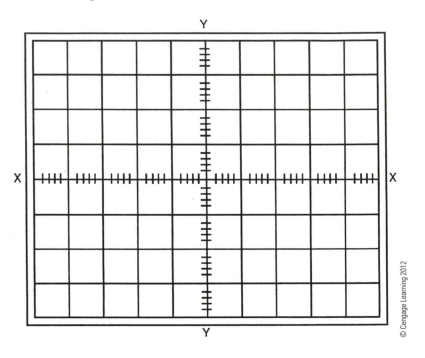

10. Open switch S_1.

11. Connect the circuit shown in Exp 22-2.

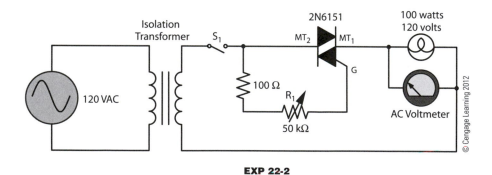

EXP 22-2

12. Adjust R_1 for maximum resistance.

13. Close switch S_1.

14. Slowly adjust resistor R_1 for 0 Ω and back to maximum resistance while observing the voltmeter connected across the load resistance.

15. Open switch S_1.

16. Connect an oscilloscope across the load resistance. Repeat steps #13 through #15.

LAB EXERCISE 23

PHASE SHIFTING THE TRIAC

1. Connect the circuit shown in Exp 23-1.

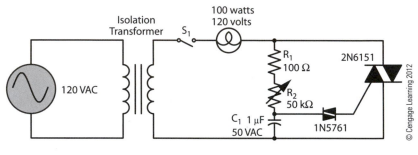

EXP 23-1

2. Adjust resistor R_2 for a maximum resistance.

3. Close switch S_1.

4. Slowly adjust resistor R_1 for 0 Ω and back to maximum resistance while observing the AC voltmeter.

5. Open switch S_1.

6. Connect an oscilloscope across the 100-watt bulb used as a load resistor. Repeat steps #3 through #5.

LAB EXERCISE 24   AC VOLTAGE CONTROL

© Shutterstock 2012

1. Connect the circuit shown in Exp 24-1.

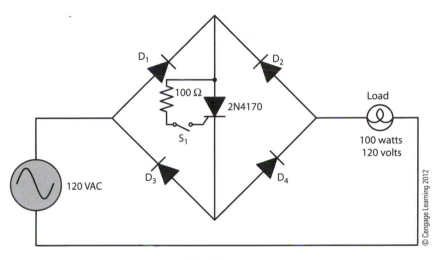

© Cengage Learning 2012

EXP 24-1

2. Connect an oscilloscope across the load. Observe the waveform when switch S_1 is closed.

3. Connect the circuit shown in Exp 24-2.

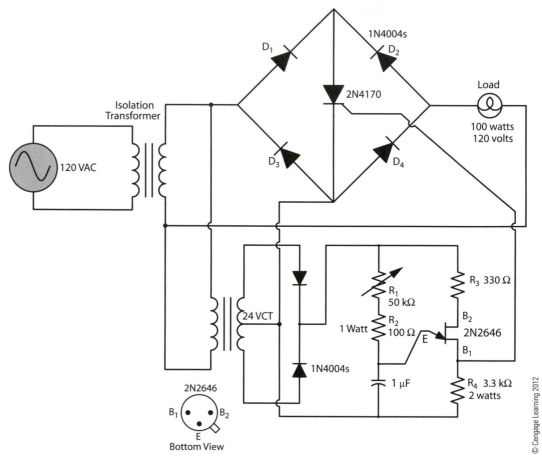

EXP 24-2

4. Connect an oscilloscope across the load. Observe the waveform when resistor R_1 is changed from one end of its range to the other.

_____ Volts   **5.** Connect an AC voltmeter across the load. Measure the voltage when resistor R_1 has been adjusted to produce the least amount of voltage across the load.

6. Slowly adjust resistor R_1 to produce the maximum voltage across the load resistor. Observe the voltmeter while making this change.

_____ Volts   **7.** Measure the AC voltage at this point.

8. Connect the circuit shown in Exp 24-3. (Note: The transistor must be heat sinked and thermal compound used.)

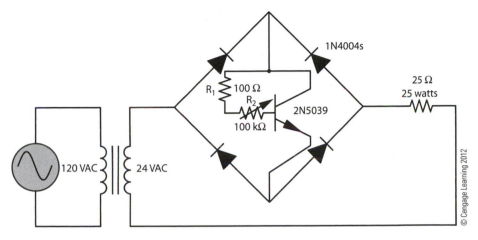

EXP 24-3

9. Connect an oscilloscope across the load resistor. Observe the waveform when resistor R_2 is adjusted from its maximum setting to its minimum setting.

10. Connect an AC voltmeter across the load.

11. Record the voltage applied to the load when resistor R_2 has been adjusted to produce the minimum voltage across the load.   _____ Volts

12. Slowly adjust resistor R_2 to produce the maximum AC voltage across the load. Observe the voltmeter when making this adjustment.

13. Measure the maximum voltage applied to the load.   _____ Volts

LAB EXERCISE 25   SOLID-STATE RELAY

1. Connect the circuit shown in Exp 25-1.

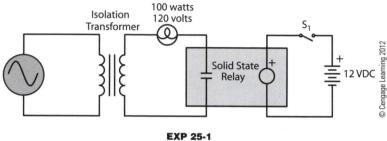

EXP 25-1

2. Connect an oscilloscope across the load resistor. Observe the waveform when switch S_1 is closed.

3. If possible, reduce the input DC voltage until the relay turns off. Measure the voltage at this point.

LAB EXERCISE 26   OSCILLATORS

© Shutterstock 2012

1. Connect the circuit shown in Exp 26-1.

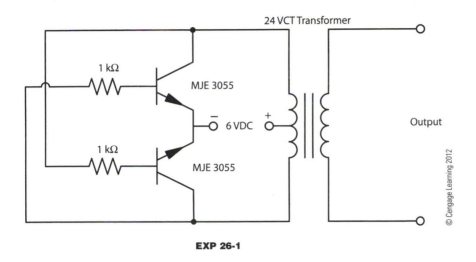

EXP 26-1

© Cengage Learning 2012

(Note: The transformer shown is a 24-volt center-tapped transformer. The secondary of the transformer is used as the primary in this circuit. The transformer primary is used as the secondary, so the transformer will become a step-up type.)

▶ CAUTION: When operating this circuit, use care because the output voltage will be high.

It is better to use a regulated DC power supply with current limit to operate this circuit. If unavailable, the input DC must be filtered or provided by a battery.)

2. Connect an oscilloscope to the output of the transformer.

3. Turn on the DC power. Record the peak-to-peak voltage and the frequency shown on the oscilloscope.

Vpp _____

F _____ Hz

4. If possible, increase the DC input voltage to 9 VDC. Record the peak-to-peak voltage and frequency shown on the oscilloscope.

Vpp _____

F _____ Hz

5. Turn off the DC power.

LAB EXERCISE 27   VOLTAGE DOUBLER

1. Connect the circuit shown in Exp 27-1.

(Note: The transformer used in the experiment is the secondary of a 24-volt center-tapped transformer. The primary side of the transformer is not used, so these leads are not connected and must remain separated.)

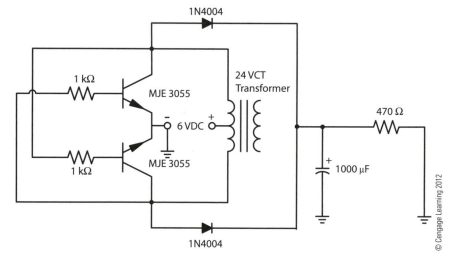

EXP 27-1

2. Connect a voltmeter and an oscilloscope across the load resistor.

3. Turn on the DC power supply. Record the voltage indicated by the voltmeter. Observe the waveform produced on the screen of the oscilloscope.

4. Turn off the power supply.

LAB EXERCISE 28   OFF-DELAY TIMER

1. Connect the circuit shown in Exp 28-1 by using the different values of capacitance shown in the chart.

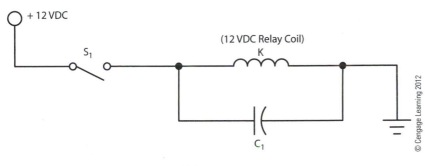

EXP 28-1

2. Fill in the chart by measuring the delay time of the relay when switch S_1 is opened for each of the capacitors listed.

CAPACITANCE	TIME
10 µF	
50 µF	
100 µF	
500 µF	
1000 µF	

3. Connect the circuit shown in Exp 28-2 by using the resistor and capacitor values shown in the chart following question 4 for R_t and C_t.

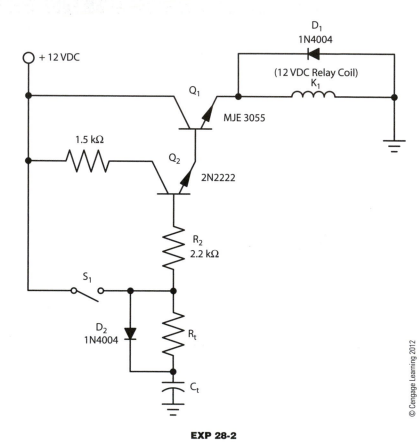

EXP 28-2

© Cengage Learning 2012

4. Fill in the chart by measuring the delay time of K_1 when S_1 is opened.

CAPACITOR	RESISTORS			
	1 kΩ	10 kΩ	100 kΩ	1 MΩ
10 µF				
50 µF				
100 µF				

LAB EXERCISE 29

ON-DELAY TIMER

© Shutterstock 2012

1. Connect the circuit shown in Exp 29-1 by using the values shown in the chart below.

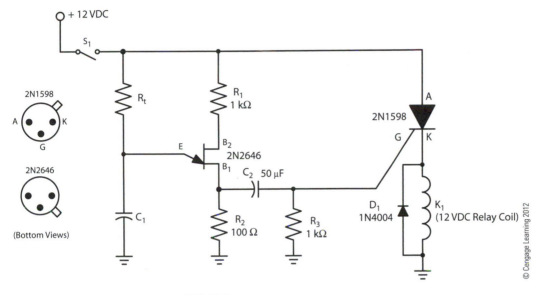

EXP 29-1

2. Fill in the chart by measuring the delay time when switch S_1 is closed. Measure the time for each capacitor and resistor value shown.

(R_t) RESISTOR	(C_1) 5 µF	(C_1) 10 µF
1 kΩ		
10 kΩ		
47 kΩ		
100 kΩ		
470 kΩ		

LAB EXERCISE 30   PULSE TIMER

1. Connect the circuit shown in Figure Exp 30-1.

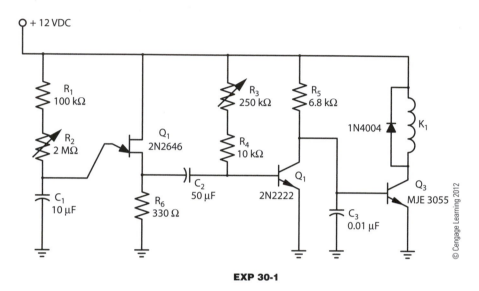

EXP 30-1

2. Vary the adjustments of resistors R_2 and R_3. Observe the operation of the circuit.

LAB EXERCISE 31

555 TIMER

(Note: In the first circuit, the 555 timer is to be operated in the monostable mode. This means that the timer must be retriggered by momentarily touching pin #2 to ground after each time sequence.)

1. Connect the circuit shown in Exp 31-1 by substituting the different values of resistance and capacitance shown in the table for R_2 and C_1.

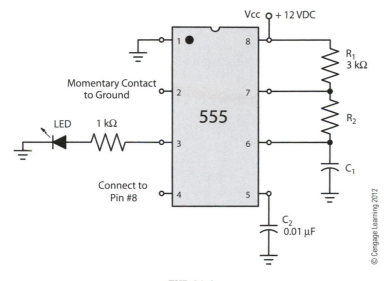

EXP 31-1

2. Observe the LED connected to the output of the timer. Fill in the chart by measuring the amount of time the output of the timer is turned on when pin #2 is pulsed to ground.

C_1	100 kΩ	220 kΩ	470 kΩ	1 MΩ
1 μF				
5 μF				
10 μF				

3. Readjust the voltage of Vcc of the figure in Laboratory Exercise 1 to 8 volts.

4. Fill in the chart by following the same procedure as in step #2.

C_1	100 kΩ	220 kΩ	470 kΩ	1 MΩ
1 μF				
5 μF				
10 μF				

(Note: In the next circuit, the 555 timer is connected in the astable mode. This means that the timer will retrigger itself at the end of each time cycle. This is done by connecting pin #2 to pin #6. Because the timer is triggered by connecting pin #2 to voltage that is one-third or less of Vcc, each time pin #7 discharges the capacitor, it also retriggers pin #2.)

5. Connect the circuit shown in Exp 31-2 by substituting the values of R_2 and C_1 shown in the chart.

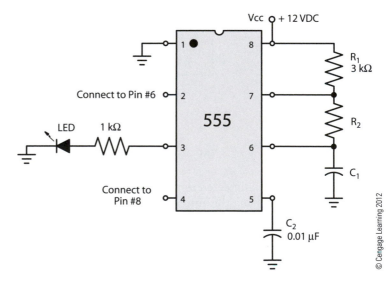

EXP 31-2

LAB EXERCISE 32 555 TIMER OSCILLATOR

1. Connect the circuit shown in Exp 32-1.

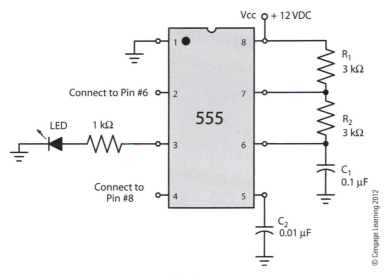

EXP 32-1

2. Connect an oscilloscope to the output at pin #3.

3. Measure the frequency of the waveform: F = 1/time.

4. Remove capacitor C_2 from the circuit. Connect a 10-k variable resistor from pin #5 to Vcc.

5. Slowly change the resistance of the variable resistor. Observe the waveform displayed on the oscilloscope.

6. Reconnect the variable resistor from pin #5 to ground.

7. Slowly change the resistance of the variable resistor. Observe the waveform displayed on the oscilloscope.

8. Adjust the variable resistor until the on time and the off time are the same.

9. Measure the frequency of the waveform.   F = _____

LAB EXERCISE 33   555 ON-DELAY TIMER

1. Connect the circuit shown in Exp 33-1 by substituting the values in the chart for resistor #1.

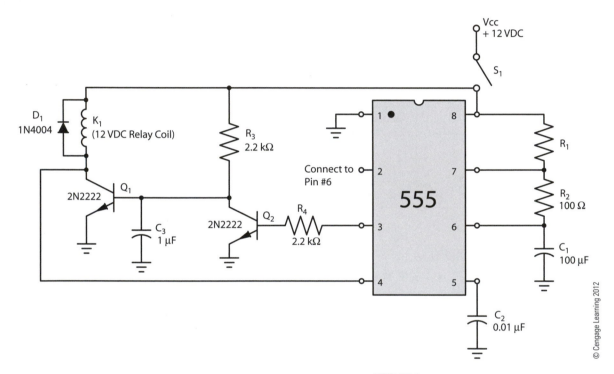

EXP 33-1

2. Fill in the chart by measuring the amount of delay for each value of resistance shown.

4.7 kΩ	10 kΩ	47 kΩ	68 kΩ	100 kΩ

LAB EXERCISE 34

555 PULSE TIMER

1. Connect the circuit shown in Exp 34-1.

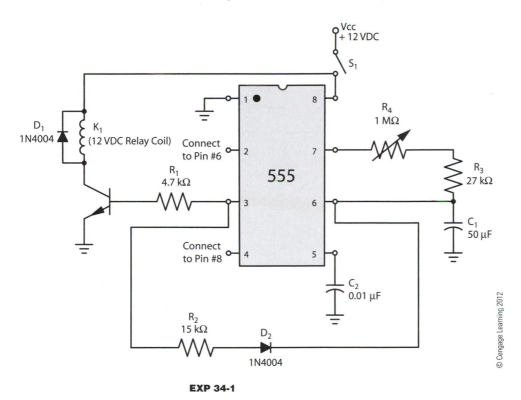

EXP 34-1

2. Adjust the setting of resistor R_4 to change the length of time between pulses.

3. Replace resistor R_2 with a 4.7 kΩ resistor and measure the length of time that relay K_1 is turned on. Then replace resistor R_2 with a 30 kΩ resistor and again measure the length of time that relay K_1 remains on.

LAB EXERCISE 35   ABOVE- AND BELOWGROUND POWER SUPPLY

1. Connect the circuit shown in Exp 35-1.

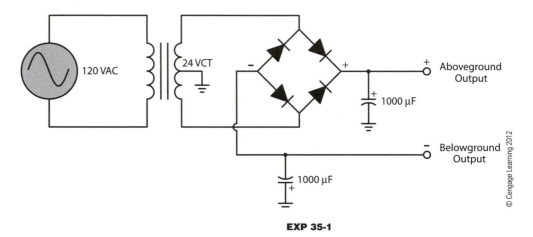

EXP 35-1

2. Using a DC voltmeter, measure the voltage between the positive output of the rectifier and ground.

3. Measure the voltage between the negative output of the rectifier and ground.

4. Measure the voltage between the positive and the negative output terminals of the rectifier.

© Shutterstock 2012

LAB EXERCISE 36   OPERATIONAL AMPLIFIER

(Note: In this lab exercise a dual-trace oscilloscope is used if available. If not available, a high impedance AC voltmeter is used to measure input and output voltages. This circuit also requires a low-voltage AC source, which is used as the input signal. A signal generator provides the best input signal, but if not available, connect the circuit shown in the Exp 36-1.)

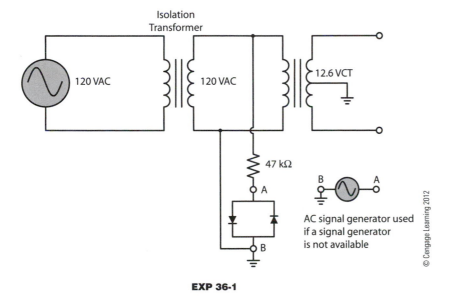

EXP 36-1

(Note: The AC signal source is taken from point A. The AC voltage at point A is the forward voltage drop of the two diodes, which is about 1.4 volts peak to peak. Be sure to connect point B back to ground [the center tap] of the transformer.)

▶ CAUTION: The 47-kΩ resistor has a voltage drop of 120 volts AC across it.

1. Connect the circuit shown in Exp 36-2.

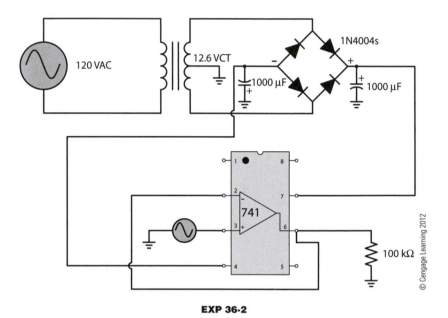

EXP 36-2

2. Using a dual-trace oscilloscope, connect one input lead to the output of the amplifier and one lead to the AC signal input.

_____ Volts   **3.** Measure the peak-to-peak voltage of the AC input signal.

_____ Volts   **4.** Measure the peak-to-peak voltage of the output of the amplifier.

V_{out}/V_{in} = _____   **5.** Compute the gain by dividing the output volts by the input volts.

_____   **6.** Observe the two voltages shown on the display. Is the output voltage inverted or noninverted when compared to the input voltage?

7. Connect the circuit shown in Exp 36-3. This schematic does not show the actual power supply, but shows only the positive and negative input voltages to the op amp.

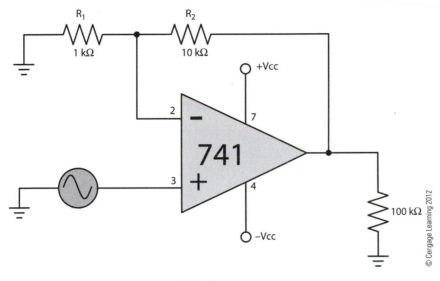

EXP 36-3

8. Calculate the gain of the amplifier by $\dfrac{R_1 + R_2}{R_1}$

Gain _____

9. Connect one oscilloscope input lead to the AC signal input. Connect the other oscilloscope lead to the output of the op amp.

10. Measure the peak-to-peak value of the AC input signal voltage.

_____ Volts

11. Measure the peak-to-peak voltage at the output of the op amp.

_____ Volts

12. Compute the gain of the amplifier by dividing the output voltage by the input voltage.

$V_{out}/V_{in} =$ _____

13. Compare this value with the value you computed in step #8.

14. Observe the voltage waveforms shown on the display of the oscilloscope. Is the output voltage inverted or noninverted when compared to the input voltage?

_____

15. Replace resistor R_1 with a 2-kΩ resistor.

16. Compute the gain of the amplifier by $\dfrac{R_1 + R_2}{R_1}$

Gain _____

17. Measure the peak-to-peak value of the AC input voltage.

_____ Volts

18. Measure the peak-to-peak value of the output voltage.

_____ Volts

19. Compute the gain of the amplifier by dividing the output voltage by the input voltage.

$V_{out}/V_{in} =$ _____

20. Compare the gain computed in step #19 to the gain computed in step #16.

21. Replace resistor R_1 with a 4.7-kΩ resistor.

Gain _____

22. Compute the gain of the circuit by $\dfrac{R_1 + R_2}{R_1}$

_____ Volts

23. Measure the peak-to-peak voltage at the output of the amplifier.

_____ Volts

24. Measure the peak-to-peak voltage of the AC input signal.

$V_{out}/V_{in} =$ _____

25. Compute the gain of the amplifier by dividing the output voltage by the input voltage.

26. Compare the gain computed in step #25 with the gain computed in step #22.

27. Connect the circuit shown in Exp 36-4.

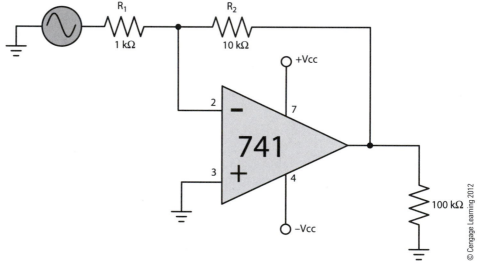

EXP 36-4

© Cengage Learning 2012

Gain _____

28. Compute the gain by R_2/R_1.

29. Connect one of the oscilloscope input leads to the AC signal source and the other lead to the output of the amplifier.

_____ Volts

30. Measure the peak-to-peak voltage of the AC signal source.

_____ Volts

31. Measure the peak-to-peak voltage of the output of the amplifier.

$V_{out}/V_{in} =$ _____

32. Compute the gain of the amplifier by dividing the output voltage by the input voltage.

33. Compare the gain computed in step #32 with the gain computed in step #28.

34. Observe the two voltages displayed on the oscilloscope. Is the output voltage inverted or noninverted when compared to the input voltage?

35. Replace resistor R_1 with a 2-kΩ resistor.

36. Compute the gain of the circuit by R_2/R_1.

Gain _____

37. Measure the peak-to-peak voltage of the input signal.

_____ Volts

38. Measure the output peak-to-peak voltage.

_____ Volts

39. Compute the gain of the amplifier by dividing the output voltage by the input voltage.

$V_{out}/V_{in} =$ _____

40. Replace resistor R_1 with a 4.7-kΩ resistor.

41. Compute the gain of the circuit by R_2/R_1.

Gain _____

42. Measure the peak-to-peak voltage of the output of the amplifer.

_____ Volts

43. Measure the peak-to-peak voltage of the AC input voltage.

_____ Volts

44. Compute the gain of the circuit by dividing the output voltage by the input voltage.

$V_{out}/V_{in} =$ _____

45. Compare the gain computed in step #44 with the gain computed in step #41.

LAB EXERCISE 37

OP AMP LEVEL DETECTOR

1. Connect the circuit shown in Exp 37-1.

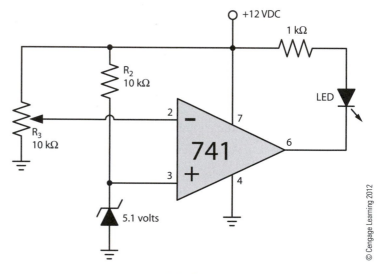

EXP 37-1

2. If available, connect a digital voltmeter (DVM) to the inverting input terminal of the op amp. Adjust R_3 until *zero volt* appears at the terminal.

_____ Volts 3. Measure the reference voltage provided by the zener diode at the noninverting input.

_____ Volts 4. Reconnect the DVM to the inverting input. Slowly adjust resistor R_3 until the LED indicates the op amp has changed from a high output to a low output. Measure the voltage at the inverting input when this change happens.

5. Connect the circuit shown in Exp 37-2.

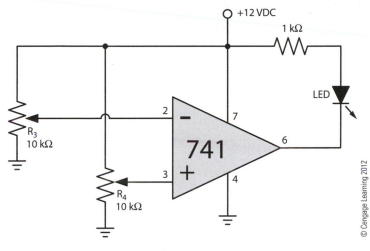

EXP 37-2

6. Connect the DVM to the noninverting input. Adjust resistor R_4 until a voltage of 3 volts is applied to the input.

7. Connect the DVM to the inverting input. Adjust resistor R_3 until a voltage of *zero volt* appears at the input.

8. Slowly adjust resistor R_3 until the LED indicates the output of the op amp has changed from its high state to its low state. Measure the voltage applied to the inverting input when this change takes place.

_____ Volts

9. Adjust resistor R_3 until *zero volt* is applied to the inverting input.

10. Connect the DVM to the noninverting input. Adjust resistor R_4 until 8 volts is applied to the input.

11. Connect the DVM to the inverting input. Slowly adjust resistor R_3 until the LED indicates the output has changed from a high state to a low state. Measure the voltage applied to the inverting input when this change occurs.

_____ Volts

12. Connect the circuit shown in Exp 37-3.

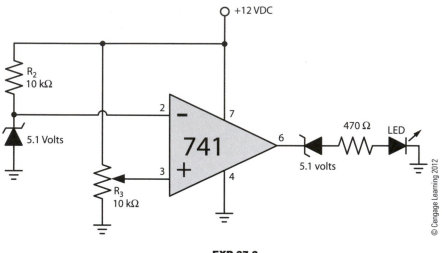

EXP 37-3

13. Adjust resistor R_3 until *zero volt* is applied to the noninverting input.

_____ Volts   **14.** Measure the voltage supplied to the inverting input by the zener diode.

_____ Volts   **15.** Connect the DVM to the noninverting input. Adjust resistor R_3 until the LED indicates the op amp has changed from a low state to a high state. Measure the input voltage when this change occurs.

16. Connect the circuit shown in Exp 37-4.

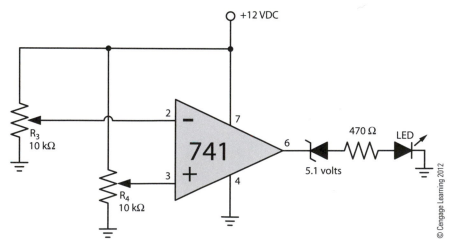

EXP 37-4

17. Adjust resistor R_3 to produce a voltage of 4 volts at the inverting input.

18. Connect the DVM to the noninverting input. Adjust resistor R_4 until the LED indicates the output of the op amp has changed state. Measure the input voltage when this change takes place.

_____ Volts

19. Readjust resistor R_3 to produce a voltage of 7 volts at the inverting input.

20. Adjust resistor R_4 until the LED indicates the output of the op amp has changed from a low state to a high state. Measure the input voltage when this change occurs.

_____ Volts

LAB EXERCISE 38   OP AMP OSCILLATOR

1. Connect the circuit shown in Exp 38-1.

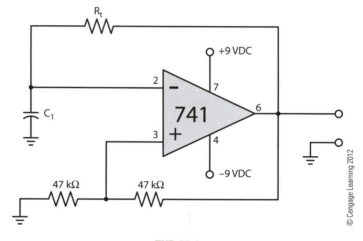

EXP 38-1

2. Connect an oscilloscope to the output of the oscillator.

3. Fill in the chart by measuring the frequency of the output waveform for each combination of values for C_t and R_t (F = 1/T).

	R_t		
C_t	100 kΩ	47 kΩ	10 kΩ
0.01 μF			
0.1 μF			

4. Connect the circuit shown in Exp 38-2.

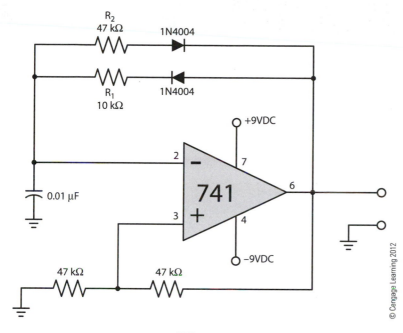

EXP 38-2

5. Connect an oscilloscope to the output of the pulse generator.

6. Measure the length of time the output is high and low.

_____ High

_____ Low

7. Find the ratio of time the output is low compared to high by dividing the low time by the high time.

Low/High = _____

8. Find the ratio of the timing resistors by dividing the value of R_2 by R_1.

R_2/R_1 = _____

9. Compare the ratio found in step #7 with that found in step #8.

10. Replace resistor R_2 with a 100-kΩ resistor.

11. Measure the amount of time the output is high and low.

_____ High

_____ Low

12. Find the ratio of the time the output is low compared to high by dividing the low time by the high time.

Low/High = _____

13. Find the ratio of the timing resistors by dividing the value of resistor R_2 by resistor R_1.

R_1/R_2 = _____

14. Compare the ratio found in step #12 with the ratio found in step #13.

LAB EXERCISE 39 VOLTAGE REGULATOR

1. Connect the circuit shown in Exp 39-1.

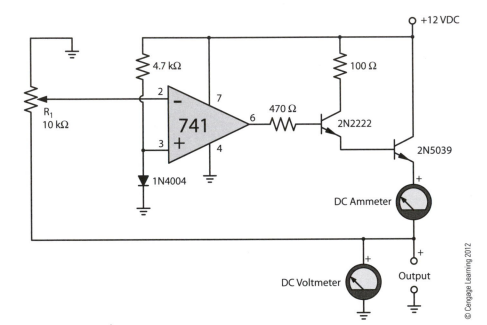

EXP 39-1

2. Adjust the value of resistor R_1 until a voltage of 4 volts appears across the output of the power supply.

_____ Volts

_____ A

3. Connect a 10-Ω, 10-watt resistor across the output of the power supply. Record the voltage and current.

4. To test the regulation of the power supply, alternately connect and disconnect the 10-Ω load resistor while observing the output voltage.

5. Disconnect the 10-Ω load resistor from the circuit. Adjust resistor R_1 until a voltage of 8 volts appears at the output.

6. Connect the 10-Ω resistor to the output of the power supply. Record the voltage and current.

_____ Volts

_____ A

7. Check the regulation of the power supply by alternately connecting and disconnecting the 10-Ω load resistor.

8. Connect the circuit shown in Exp 39-2.

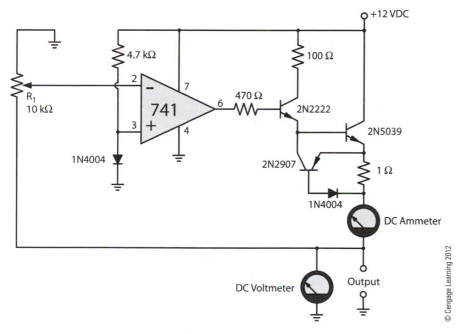

EXP 39-2

9. Adjust resistor R_1 until a voltage of 4 volts appears at the output of the power supply.

10. Connect a 10-Ω resistor across the output of the power supply. Record the voltage and current.

_____ Volts

_____ A

11. Check the voltage regulation by alternately connecting and disconnecting the 10-Ω resistor to the output of the power supply.

12. Disconnect the 10-Ω load resistor from the output of the power supply. Adjust the output of the power supply for a voltage of 8 volts.

_____ Volts

_____ A

13. Connect the 10-Ω load resistor to the output of the power supply. Record the voltage and current.

14. Disconnect the 10-Ω load resistor.

_____ A

15. To test the short-circuit protection of the power supply, momentarily short-circuit the output of the power supply. When the output is shorted, the voltmeter should drop to 0 volt. When the short is removed, the voltmeter should return to the original setting of the output. Record the current when the output is shorted. This is the maximum current the power supply will deliver before going into current limit.

16. Connect a 2-Ω resistor in parallel with the 1-Ω sense resistor as shown in Exp 39-3. (Note: Two 1-Ω resistors may have to be connected in series.)

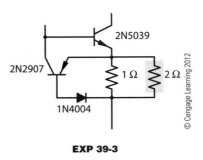

EXP 39-3

_____ A

17. Repeat the test in step #15 to test the maximum output current of the power supply.

LAB EXERCISE 40

1. Using the diagram of a 7408N quad two-input AND gate shown in Exp 40-1, connect pin #14 to +5 volts and pin #7 to ground.

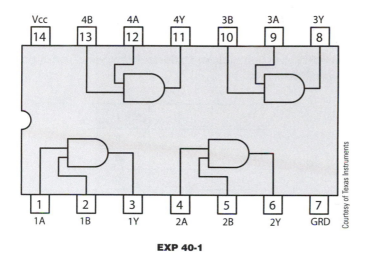

EXP 40-1

2. Use an LED connected in series with a 330-Ω resistor, as shown in Exp 40-2, to indicate when the output of the gate is high or low.

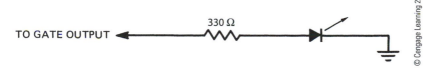

EXP 40-2

3. Fill in the truth table for an AND gate by connecting an input to ground when a 0 is indicated, or to +5 volts when a 1 is indicated.

A	B	Y
0	0	
0	1	
1	0	
1	1	

4. Compare the truth table in Laboratory Exercise 3 with that shown below.

A	B	Y
0	0	0
0	1	0
1	0	0
1	1	1

Truth table for a two-input AND gate

5. Use the base diagram shown in Exp 40-3 to connect a 7432N quad two-input OR gate.

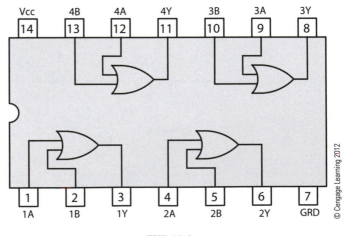

© Cengage Learning 2012

EXP 40-3

6. Fill in the truth table for an OR gate by using the same procedure as outlined in steps #2 and #3.

A	B	Y
0	0	
0	1	
1	0	
1	1	

7. Compare the truth table shown in Laboratory Exercise 6 with that shown below.

A	B	Y
0	0	0
0	1	1
1	0	1
1	1	1

Truth table for a two-input OR gate

8. Use the diagram shown in Exp 40-4 to connect a 7402N quad two-input NOR gate.

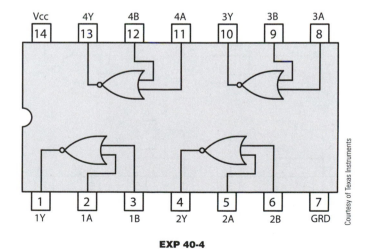

EXP 40-4

9. Fill in the truth table for a NOR gate.

A	B	Y
0	0	
0	1	
1	0	
1	1	

10. Compare the truth table shown in Laboratory Exercise 9 with that shown below.

A	B	Y
0	0	1
0	1	0
1	0	0
1	1	0

Truth table for a two-input NOR gate

11. Use the diagram shown in Exp 40-5 to connect a 7400N quad two-input NAND gate.

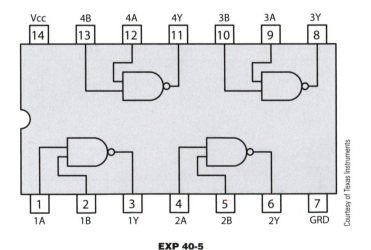

EXP 40-5

Courtesy of Texas Instruments

12. Fill in the truth table for an NAND gate.

A	B	Y
0	0	
0	1	
1	0	
1	1	

13. Compare the truth table shown in Laboratory Exercise 12 with that shown below.

A	B	Y
0	0	1
0	1	1
1	0	1
1	1	0

Truth table of a two-input NAND gate

14. Use the diagram shown in Exp 40-6 to connect a 7404N hex INVERTER.

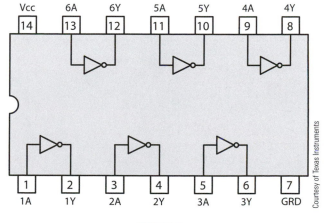

EXP 40-6

15. Fill in the truth table below for an INVERTER.

A	Y
0	
1	

16. Compare the truth table shown in Laboratory Exercise 15 with that shown below.

A	Y
0	1
1	0

Truth table for an INVERTER (NOT)

LAB EXERCISE 41

BOUNCELESS SWITCH

© Shutterstock 2012

1. Connect the circuit shown in Exp 41-1.

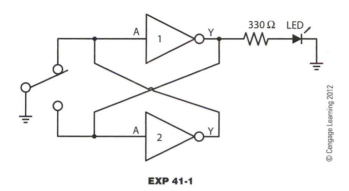

EXP 41-1

© Cengage Learning 2012

2. Change the position of the switch several times and observe the LED.

3. Connect the circuit shown in Exp 41-2.

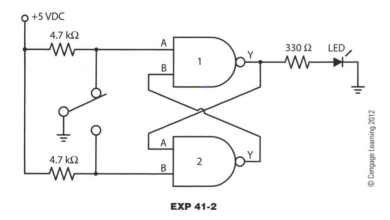

EXP 41-2

© Cengage Learning 2012

4. Change the position of the switch several times and observe the LED.

© Shutterstock 2012

1. Connect the circuit shown in Exp 42-1.

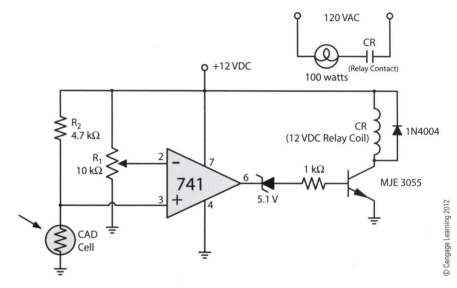

EXP 42-1

© Cengage Learning 2012

2. Using a dark object, such as a piece of dark-colored paper or cloth, cover the cad cell.

3. Adjust resistor R_1 until you reach a point that a slight movement of the variable resistor will cause the light to turn on or off. When this point is reached, turn the resistor about one-eighth of a turn in the direction that causes the light to remain turned on.

4. Remove the dark cover from the cad cell. The light should turn off.

5. Replace the dark cover over the cad cell and the light should turn on.

6. Connect the circuit shown in Exp 42-2.

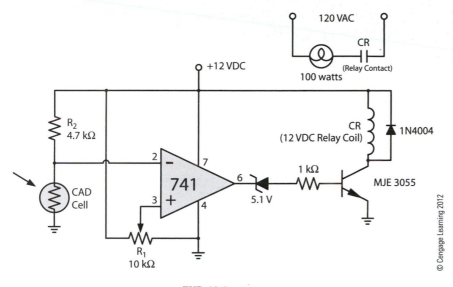

EXP 42-2

7. Place the dark cover over the cad cell and adjust resistor R_1 until a light movement of the resistor will cause the light to turn on or off. When this point is reached, turn the resistor about one-eighth of a turn in the direction that causes the light to remain turned off.

8. Remove the cover from the cad cell and the light should turn on.

9. Replace the cover and the light should turn off.

LAB EXERCISE 43   ANILMAL FEEDER

1. Connect the circuit shown in Exp 43-1.

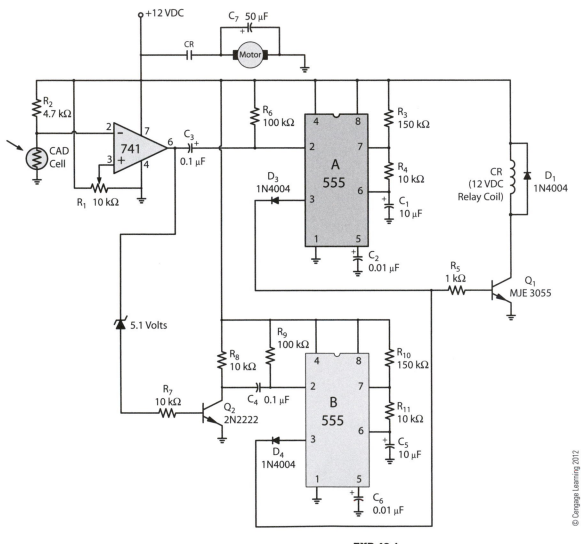

EXP 43-1

(Note: The 555 timers have been designed for a delay of about three seconds each. The 150-kΩ resistors can be changed if longer or shorter delays are desired. For a longer delay, increase the value of resistor R_3 or R_{10}. For a shorter delay, decrease the value of resistor R_3 or R_{10}.

   If a small DC motor is not available, connect a 1-k resistor in series with an LED, as shown in the figure, to represent the motor.)

2. Set the sensitivity of the photodetector as described in the previous unit.

3. Test the operation of the circuit as described in this unit.

LAB EXERCISE 44

© Shutterstock 2012

1. Connect the circuit shown in Exp 44-1. (Note: Use a single-pole switch to represent the overload contact.)

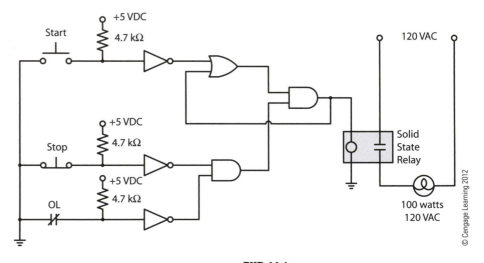

EXP 44-1

2. Test the operation of the circuit as described in this unit.

LAB EXERCISE 45 — ELECTRONIC LOCK

1. Use the connection diagrams shown below. Label the schematic shown in Exp 45-1 with the proper IC pin numbers.

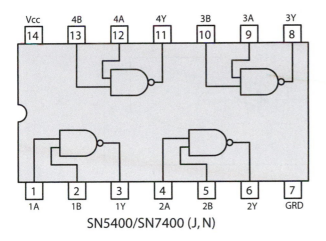

SN5400/SN7400 (J, N)

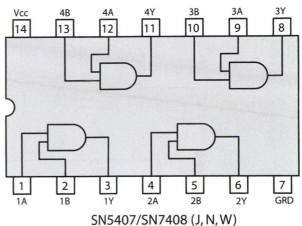

SN5407/SN7408 (J, N, W)

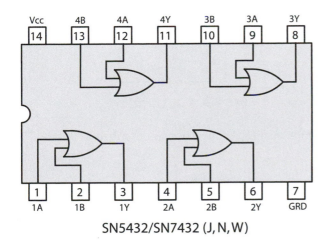

SN5432/SN7432 (J, N, W)

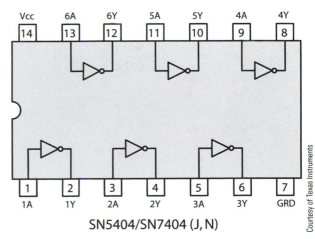

SN5404/SN7404 (J, N)

EXP 45-1

2. Connect the circuit shown in Exp 45-2.

(Note: If enough normally open push buttons are not available, use a wire connected to ground. The wire will act as a push button if momentarily touched to the proper point in the circuit.)

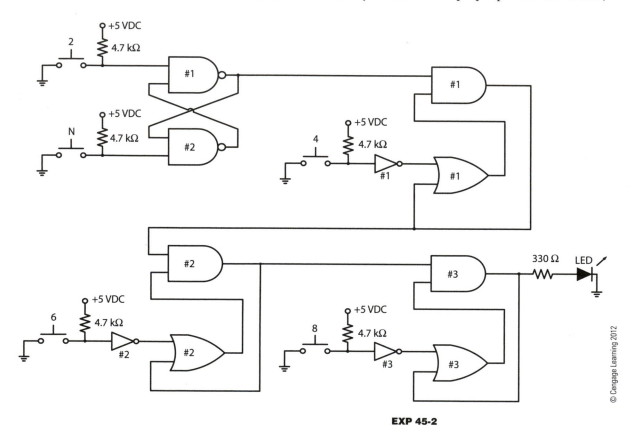

EXP 45-2

3. Test the circuit for the proper operation as described in this unit.

LAB EXERCISE 46

SOLID-STATE THERMOSTAT

© Shutterstock 2012

1. Connect the circuit shown in Exp 46-1.

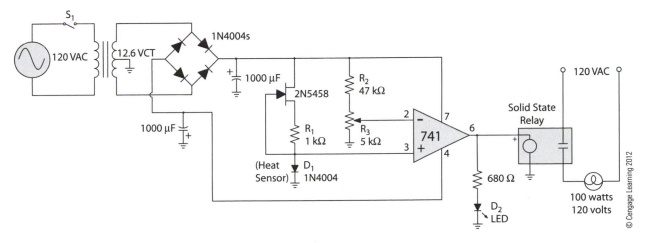

EXP 46-1

© Cengage Learning 2012

2. Close switch S_1 and adjust the variable resistor until the output of the op amp just turns on.

3. Hold the heat-sensing diode between two fingers. (The heat of your body should raise the diode's temperature enough to cause the output to turn off.)

4. When the output turns off, release the heat-sensing diode and permit it to cool. This should cause the output to turn on again.

5. Open switch S_1.

6. Change the connections to the inverting and noninverting inputs as shown in Figure 48-4. This should change the operation of the thermostat from a heating mode to a cooling mode.

7. Turn on Switch S_1.

8. Adjust the variable resistor until the output just turns off.

9. Hold the heat-sensing diode between two fingers. This should cause the output to turn on.

10. Release the diode and permit the diode to cool. The output should turn off again.

11. Open switch S_1.

12. Disconnect the circuit and return the components to their proper place.

APPENDIX A: TESTING SOLID-STATE COMPONENTS

PROCEDURE 1: TESTING A DIODE

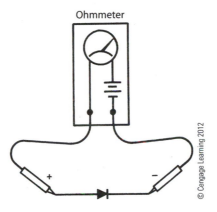

Step 1. Connect the ohmmeter leads to the diode. Notice whether the meter indicates continuity through the diode.

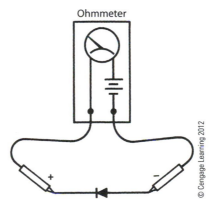

Step 2. Reverse the diode connection to the ohmmeter. Notice whether the meter indicates continuity through the diode. The ohmmeter should indicate continuity through the diode in only one direction. (Note: If continuity is not indicated in either direction, the diode is open. If continuity is indicated in both directions, the diode is shorted.)

PROCEDURE 2: TESTING A TRANSISTOR

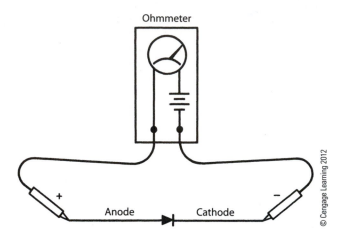

Step 1. Using a diode, determine which ohmmeter lead is positive and which is negative. The ohmmeter will indicate continuity through the diode only when the positive lead is connected to the anode of the diode and the negative lead is connected to the cathode.

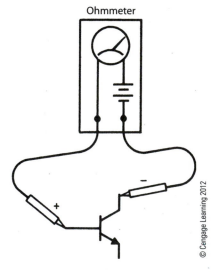

Step 2. If the transistor is an NPN, connect the positive ohmmeter lead to the base and the negative lead to the collector. The ohmmeter should indicate continuity. The reading should be about the same as the reading obtained when the diode was tested.

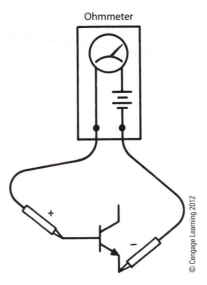

Ohmmeter

© Cengage Learning 2012

Step 3. With the positive ohmmeter lead still connected to the base of the transistor, connect the negative lead to the emitter. The ohmmeter should again indicate a forward diode junction. (Note: If the ohmmeter does not indicate continuity between the base–collector or the base–emitter, the transistor is open.)

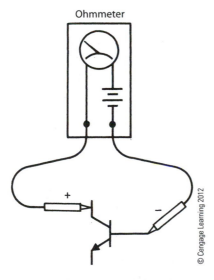

Ohmmeter

© Cengage Learning 2012

Step 4. Connect the negative ohmmeter lead to the base and the positive lead to the collector. The ohmmeter should indicate infinity or no continuity.

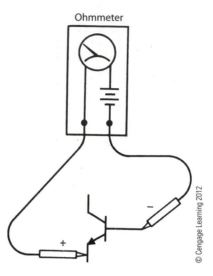

Step 5. With the negative ohmmeter lead connected to the base, reconnect the positive lead to the emitter. There should again be no indication of continuity. (Note: If a very high resistance is indicated by the ohmmeter, the transistor is "leaky" but may still operate in the circuit. If a very low resistance is seen, the transistor is shorted.)

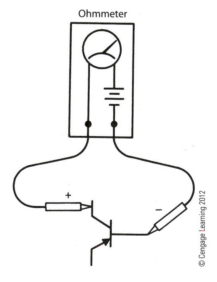

Step 6. To test a PNP transistor, reverse the polarity of the ohmmeter leads and repeat the test. If the negative ohmmeter lead is connected to the base, a forward diode junction should be indicated when the positive lead is connected to the collector or emitter.

Ohmmeter

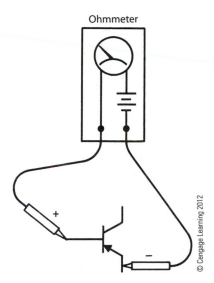

© Cengage Learning 2012

Step 7.   If the positive ohmmeter lead is connected to the base of a PNP transistor, no continuity should be indicated when the negative lead is connected to the collector or the emitter.

PROCEDURE 3: TESTING A UNIJUNCTION TRANSISTOR

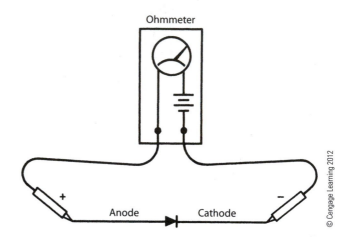

Step 1. Using a junction mode, determine which ohmmeter lead is positive and which is negative. The ohmmeter will indicate continuity when the positive lead is connected to the anode and the negative lead is connected to the cathode.

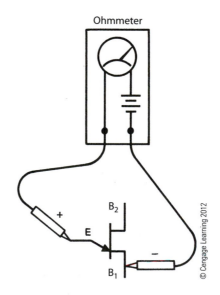

Step 2. Connect the positive ohmmeter lead to the emitter lead and the negative lead to base #1. The ohmmeter should indicate a forward diode junction.

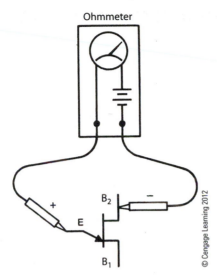

Step 3.   With the positive ohmmeter lead connected to the emitter, reconnect the negative lead to base #2. The ohmmeter should again indicate a forward diode junction.

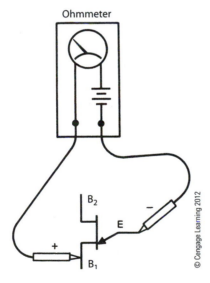

Step 4.   If the negative ohmmeter lead is connected to the emitter, no continuity should be indicated when the positive lead is connected to either base #1 or base #2.

PROCEDURE 4: TESTING AN SCR

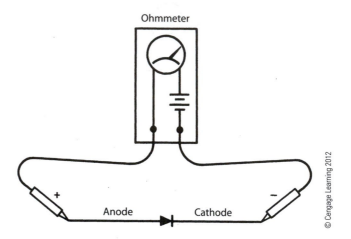

Step 1.   Using a junction diode, determine which ohmmeter lead is positive and which is negative. The ohmmeter will indicate continuity only when the positive lead is connected to the anode of the diode and the negative lead is connected to the cathode.

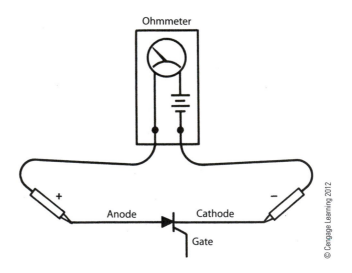

Step 2.   Connect the positive ohmmeter lead to the anode of the SCR and the negative lead to the cathode. The ohmmeter should indicate no continuity.

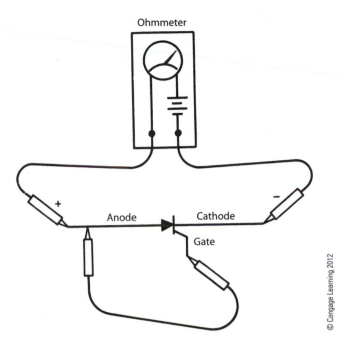

Step 3. Using a jumper lead, connect the gate of the SCR to the anode. The ohmmeter should indicate a forward diode junction when the connection is made. (Note: If the jumper is removed, the SCR may continue to conduct or it may turn off. This will be determined by whether the ohmmeter can supply enough current to keep the SCR above its holding-current level.)

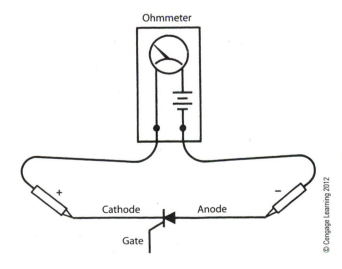

Step 4. Reconnect the SCR so that the cathode is connected to the positive ohmmeter lead and the anode is connected to the negative lead. The ohmmeter should indicate no continuity.

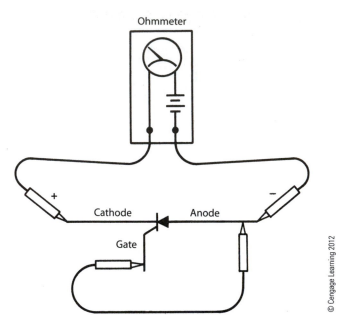

© Cengage Learning 2012

Step 5.   If a jumper lead is used to connect the gate to the anode, the ohmmeter should indicate no continuity. (Note: SCRs designed to switch large currents (50 A or more) may indicate some leakage current with this test. This is normal for some devices.)

PROCEDURE 5: TESTING A TRIAC

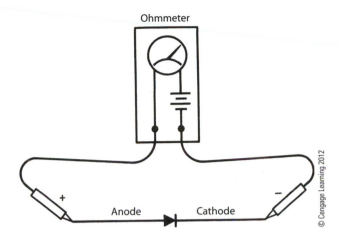

Step 1. Using a junction diode, determine which ohmmeter lead is positive and which is negative. The ohmmeter will indicate continuity only when the positive lead is connected to the anode and the negative lead is connected to the cathode.

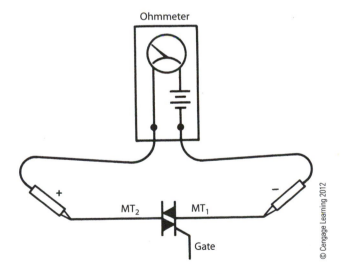

Step 2. Connect the positive ohmmeter lead to MT_2 and the negative lead to MT_1. The ohmmeter should indicate no continuity through the triac.

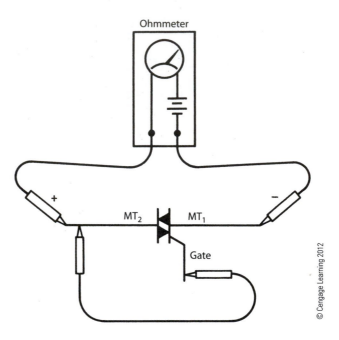

Step 3. Using a jumper lead, connect the gate of the triac to MT_2. The ohmmeter should indicate a forward diode junction.

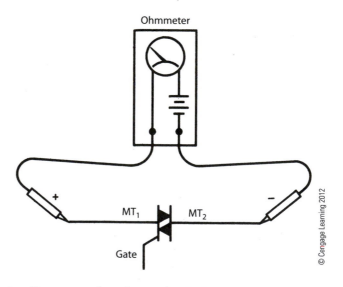

Step 4. Reconnect the triac so that MT_1 is connected to the positive ohmmeter lead and MT_2 is connected to the negative lead. The ohmmeter should indicate no continuity through the triac.

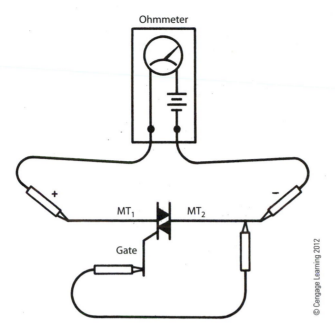

© Cengage Learning 2012

Step 5.   Using a jumper lead, again connect the gate to MT_2. The ohmmeter should indicate a forward diode junction.

APPENDIX B: OHM'S LAW FORMULAS

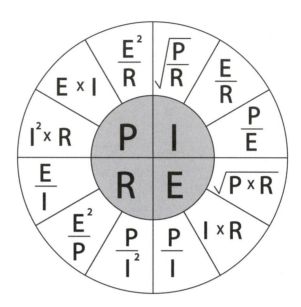

P = Power (Watts)
I = Intensity (Amperes)
R = Resistance (Ohms)
E = E.M.F. (Volts)

© Cengage Learning 2012

APPENDIX C: PARTS LIST

DIODES:	Light emitting
	1N4004 Silicon 400PRV 1 A
	12-volt zener 5 watts
	5.1-volt zener 0.25 watt
TRANSISTORS:	2N2222 NPN Silicon 0.5 watt Vce
	30 volts IC 150 mA
	2N2907 PNP Silicon 0.5 watt Vce
	30 volts IC 150 mA
	2N5039 NPN Silicon 140 watts Vce
	120 IC 10 A
	2N2646 UJT 300 mW
	MJE 3055 NPN Silicon 90 watts Vce
	60 IC 4 A
FET:	2N5458 N-Channel JFET
SCRs:	2N4170 200 volts 8 A Igate 20 mA
	Egate 1.5 volts
	2N1598 300 volts 1.6 A
DIAC:	1N5761 Turn on 32 volts Turn off
	25 volts Imax 2 A
TRIAC:	2N6151 200 volts 10 A
SOLID-STATE RELAY:	General Electric #CR120SR110D
	240 volts AC 10 A (or) International
	Rectifier #D2410
INTEGRATED CIRCUITS:	555 Timer
	741 Operational Amplifier
	7400N Quad 2-input NAND gate TTL
	7402N Quad 2-input NOR gate TTL
	7404N hex INVERTER TTL
	7408N Quad 2-input AND gate TTL
	7432N Quad 2-input OR gate TTL
TRANSFORMERS:	117/117 volts isolation transformer 2 A
	117/24 volts center-tapped 3 A
	117/12.6 volts center-tapped 3 A
RELAY:	12 volts DC Potter Brumfield or
	equivalent
RESISTORS (one-half watt):	10 Ω, 27 Ω, 100 Ω, 330 Ω, 470 Ω, 1 kΩ,
	1.2 kΩ, 1.5 kΩ, 2.2 kΩ, 3 kΩ, 3.3 kΩ,
	4.7 kΩ, 6.8 kΩ, 10 kΩ, 15 kΩ, 22 kΩ,
	27 kΩ, 47 kΩ, 68 kΩ, 75 kΩ, 100 kΩ,
	220 kΩ, 470 kΩ, 1 MΩ

RESISTORS (SPECIAL):	200 Ω 2 watts 100 Ω 2 watts 100 Ω Light bulb 3.3 kΩ 2 watts 25 Ω 25 watts
RESISTORS (VARIABLE):	1 kΩ, 5 kΩ, 10 kΩ, 50 kΩ, 250 kΩ, 1 MΩ, 2 MΩ
SWITCHES:	DPDT, SPST, Push buttons (NO and NC)
CAPACITORS (50 volt):	0.01 μF, 0.1 μF, 1 μF, 2 μF, 5 μF, 10 μF, 25 μF, 50 μF, 100 μF, 150 μF, 200 μF, 470 μF, 1000 μF
CAD CELL	Radio Shack or equivalent
PIEZO BUZZER	Sonalert #SC628 or equivalent

APPENDIX D: SCHEMATIC SYMBOLS

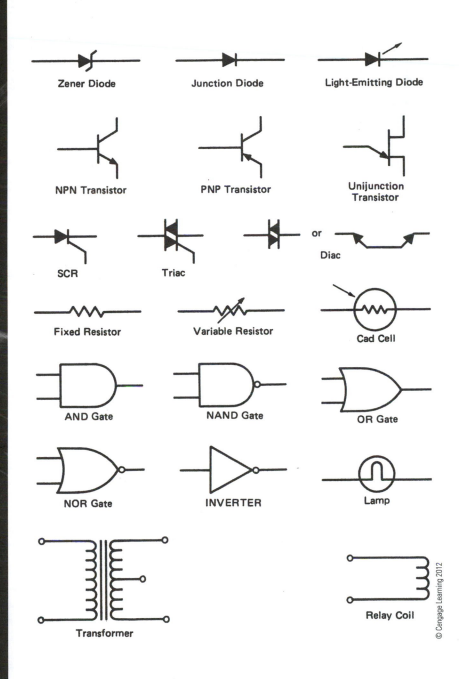

Zener Diode

Junction Diode

Light-Emitting Diode

NPN Transistor

PNP Transistor

Unijunction Transistor

SCR

Triac

or Diac

Fixed Resistor

Variable Resistor

Cad Cell

AND Gate

NAND Gate

OR Gate

NOR Gate

INVERTER

Lamp

Transformer

Relay Coil

© Cengage Learning 2012

APPENDIX E: CONVERSION FACTORS FOR SINE-WAVE VOLTAGES

TO CHANGE	TO	MULTIPLY BY
PEAK	RMS	0.707
PEAK	AVERAGE	0.637
PEAK	PEAK TO PEAK	2
RMS	PEAK	1.414
AVERAGE	PEAK	1.567
RMS	AVERAGE	0.9
AVERAGE	RMS	1.111

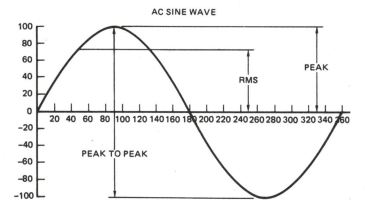

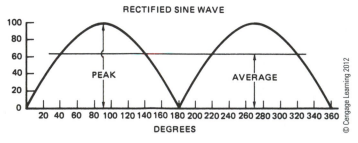

GLOSSARY

AC (alternating current) Current that reverses its direction of flow periodically. Reversals generally occur at regular intervals.

alternating current See *AC*.

alternator A machine used to generate alternating current by rotating conductors through a magnetic field.

amplifier A device used to increase a signal.

amplitude The highest value reached by a signal, voltage, or current.

AND gate A digital logic gate that must have all of its inputs high to produce an output.

anode The positive terminal of an electronic device.

applied voltage The amount of voltage connected to a circuit or device.

astable mode The state in which an oscillator can continually turn itself on and off, or continually change from positive to negative output.

attenuator A device that decreases the amount of signal, voltage, or current.

barrier charge The potential developed across a semiconductor junction.

base The semiconductor region between the collector and emitter of a transistor. The base controls the current flow through the collector–emitter circuit.

base current The amount of current that flows through the base–emitter section of a transistor.

bias A DC voltage applied to the base of a transistor to preset its operating point.

bounceless switch A circuit used to eliminate contact bounce in mechanical contacts.

bridge circuit A circuit that consists of four sections connected in series to form a closed loop.

bridge rectifier A device constructed with four diodes that converts both positive and negative cycles of AC voltage into DC voltage.

cad cell A device that changes its resistance with a change of light intensity.

capacitance The electrical size of a capacitor.

capacitive Any circuit or device having characteristics similar to those of a capacitor.

capacitor A device made with two conductive plates separated by an insulator or dielectric.

cathode The negative terminal of a device.

cathode-ray tube (CRT) An electron beam tube in which the beam of electrons can be focused to any point on the face of the tube. The electron beam causes the face of the tube to produce light when it is struck by the beam.

center-tapped transformer A transformer that has a wire connected to the electrical midpoint of its winding. Generally the secondary is tapped.

charge time The amount of time necessary to charge a capacitor.

choke An inductor designed to present an impedance to AC current, or to be used as the current filter of a DC power supply.

circuit An arrangement of electrical devices to form one or more complete paths for current flow.

clock timer A time-delay device that uses an electric clock to measure the delay period.

collapse (of a magnetic field) Occurs when a magnetic field suddenly changes from its maximum value to a zero value.

collector A semiconductor region of a transistor that must be connected to the same polarity as the base.

comparator A device or circuit that compares two like quantities such as voltage levels.

conduction level The point at which an amount of voltage or current will cause a device to conduct.

conductor A device or material that permits current to flow through it easily.

CRT See *cathode-ray tube.*

current The rate of the flow of electrons.

current flow The flow of electrons.

current generator A circuit designed to deliver a certain amount of current as opposed to a certain amount of voltage.

current rating The amount of current flow a device is designed to withstand.

dark current The current that flows in a photo detector when there is insufficient light to turn the device on.

dashpoint timer A device using a piston moving through a liquid to produce a time delay.

DC (direct current) Current that does not reverse its direction of flow.

DE-MOSFET (depletion-enhancement metal oxide semiconductor field effect transistor) This transistor can be operated in either a depletion or enhancement mode.

delta connection A circuit formed by connecting three electrical devices in series to form a closed loop. It is used most often in three-phase connections.

diac A bidirectional diode.

digital device A device that has only two states of operation, on and off.

digital logic Circuit elements connected in such a manner as to solve problems using components that have only two states of operation.

digital voltmeter A voltmeter that uses a direct reading numerical display as opposed to a meter movement.

diode A two-element device that permits current to flow through it in only one direction.

direct current See *DC.*

doping The controlled addition of impurities to a semiconductor material to achieve a desired characteristic.

E-MOSFET (enhancement metal oxide semiconductor field effect transistor) This transistor can be operated in the enhancement mode only.

emitter The semiconductor region of a transistor that must be connected to a polarity different from the base.

esaki diode A diode that exhibits a negative resistance.

EXCLUSIVE-OR gate A digital logic gate that will produce an output when its inputs have opposite states of logic level.

FET See *field effect transistor.*

field effect transistor (FET) A transistor that controls the flow of current through it with an electric field.

filter A device used to remove the ripple produced by a rectifier.

frequency The number of complete cycles of AC voltage that occur in 1 second.

gain The increase in signal power produced by an amplifier.

gate A device that has multiple inputs and a single output, or one terminal of some solid-state devices such as SCRs or triacs.

heat sink A metallic device designed to increase the surface area of an electronic component for the purpose of removing heat at a faster rate.

holding contacts Contacts used for the purpose of maintaining current flow to the coil of a relay.

holding current The amount of current needed to keep an SCR or triac turned on.

hysteresis loop A graphic curve that shows the value of magnetizing force for a particular type of material.

IGFET (insulated gate field effect transistor) This device is the same as a MOSFET.

impedance The total opposition to current flow in an electrical circuit.

induced current Current produced in a conductor by the cutting action of a magnetic field.

inductor A coil.

input voltage The amount of voltage connected to a device or circuit.

insulator A material used to electrically isolate two conductive surfaces.

internal relay Digital logic circuits in a programmable controller that can be programmed to operate in the same manner as control relays.

INVERTER (gate) A digital logic gate that has an output opposite its input.

inverting input The input terminal of an operational amplifier that will cause the output to assume the opposite polarity of voltage.

isolation transformer A transformer whose secondary winding is electrically isolated from its primary winding.

JFET (junction field effect transistor) A field effect transistor formed by combining layers of semiconductor material. The JFET has an input impedance equal to the reverse bias impedance of a pn junction. This impedance is approximately 20,000 MΩ.

junction diode A diode that is made by joining together two pieces of semiconductor material.

kick back diode A diode used to eliminate the voltage spike induced in a coil by the collapse of a magnetic field.

lattice structure An orderly arrangement of atoms in a crystalline material.

light current The amount of current that flows through a photo detector when there is sufficient light to turn it on.

LED (light-emitting diode) A diode that will produce light when current flows through it.

light-emitting diode See *LED*.

magnetic field The space in which a magnetic force exists.

microprocessor A small computer. The central processing unit is generally made from a single integrated circuit.

mode A state of condition.

monostable mode The state in which an oscillator or timer will operate through only one sequence of events.

MOSFET (metal oxide semiconductor field effect transistor) A field effect transistor that has no electrical connection between the gate and channel. The input impedance is approximately 2 billion ohms.

motor controller A device used to control the operation of a motor.

n-channel A field effect transistor so constructed that n-type semiconductor material is used for the channel through the device.

NAND gate A digital logic gate that will produce a high output only when all of its inputs are in a low state.

negative resistance The property of a device in which an increase of current flow causes an increase of conductance. The increase of conductance causes a decrease in the voltage drop across the device.

noninverting input The input of an operational amplifier that produces the same polarity of voltage at the output.

NOR gate A digital logic gate that will produce a high output when any of its inputs are low.

normally closed The contact of a relay that is closed when the relay coil is de-energized.

normally open The contact of a relay that is open when the relay coil is de-energized.

off-delay timer A timer whose contacts change position immediately when the coil or circuit is energized, but delay returning to their normal position when the coil or circuit is de-energized.

ohmmeter A meter used to measure resistance.

on-delay timer A timer whose contacts will delay changing position when the coil or circuit is energized, but change back immediately when the coil or circuit is de-energized.

op amp An operational amplifier.

operational amplifier A direct-coupled integrated circuit amplifier used for high-output current applications.

optoisolator A device used to connect sections of a circuit by means of a light beam.

oscillator A device or circuit that is used to change DC voltage into AC voltage.

oscilloscope An instrument that measures the amplitude of voltage with respect to time.

out-of-phase voltage A voltage that is not in phase when compared to some other voltage or current.

output pulse A short duration voltage or current that can be negative or positive, produced at the output of a device or circuit.

p-channel A field effect transistor so constructed that p-type semiconductor material is used for the channel through the device.

panelboard A metallic or nonmetallic panel used to mount electrical controls, devices, or equipment.

parallel circuit A circuit that has more than one path for current flow.

peak-inverse/peak-reverse voltage The rating of a semiconductor device that indicates the maximum amount of voltage in the reverse direction that can be applied to the device.

peak-to-peak voltage The amplitude of voltage measured from the negative peak of an AC waveform to the positive peak.

peak voltage The amount of voltage of a waveform measured from the zero voltage point to the positive or negative peak.

phase shift A change in the phase relationship between two quantities of voltage or current.

photodetector A device that responds to a change in light intensity.

photodiode A diode that will conduct in the presence of light and not conduct when in darkness.

pneumatic timer A device that uses the displacement of air in a bellows or diaphragm to produce a time delay.

polarity The characteristic of a device that exhibits opposite quantities within itself: positive and negative.

potentiometer A variable resistor with a sliding contact that is used as a voltage divider.

power rating The rating of a device that indicates the amount of current flow and voltage drop that can be permitted.

pressure switch A device that senses the presence or absence and causes a set of contacts to open or close.

pulse generator An oscillator that produces a voltage of short duration on regular intervals.

RC time constant The time constant of a resistor and capacitor connected in series. The time in secondary is equal to the resistance in ohms multiplied by the capacitance in farads.

reactance The opposition to current flow in an AC circuit offered by pure inductance or pure capacitance.

rectifier A device or circuit used to change AC voltage into DC voltage.

regulator A device that maintains a quantity at predetermined level.

relay A magnetically operated switch that may have one or more sets of contacts.

resistance The opposition to current flow in an AC or a DC circuit.

resistive temperature detector (RTD) A device that changes its resistance with a change in temperature. These devices are made of metal and exhibit a positive temperature coefficient.

ripple The AC component in the output of a DC power supply caused by improper filtering.

RMS value The value of AC voltage that will produce as much power when connected across a resistor as a like amount of DC voltage.

root-mean-square value See *rms value*.

RTD See *resistive temperature detector*.

saturation The maximum amount of magnetic flux a material can hold.

schematic An electrical diagram that shows components in their electrical sequence without regard for physical location.

SCR (silicon-controlled rectifier) A four-layer semiconductor device that is a rectifier and must be triggered by a pulse applied to the gate before it will conduct.

semiconductor A material that contains four valence electrons and is used in the production of solid-state devices. The most common semiconductors are silicon and germanium.

series aiding Two or more voltage-producing devices connected in series in such a manner that their voltages add to produce a higher total voltage.

series circuit An electric circuit formed by the connection of one or more components in such a manner that there is only one path for current flow.

shockley diode A four-layer diode used as a trigger or switching device.

signal generator A test instrument used to produce a low value AC voltage for the purpose of testing or calibrating electronic equipment.

silicon-controlled rectifier See *SCR*.

silicon unilateral switch (SUS) A four-layer device similar to an SCR, except that a zener junction is added to the anode gate, which permits it to be triggered into conduction at approximately 8 volts. The SUS can also be triggered by a negative pulse applied to the gate.

sine-wave voltage A voltage waveform whose value at any point is proportional to the trigonometric sine of the angle of the generator producing it.

solenoid A magnetic device used to convert electrical energy into linear motion.

solenoid valve A value operated by an electric solenoid. See *solenoid*.

solid-state electronic device An electronic component constructed from semiconductor material.

stealer transistor A transistor used in such a manner as to force some other component to remain in the off state by shutting its current to electrical ground.

step-down transformer A transformer that produces a lower voltage at its secondary than is applied to its primary.

step-up transformer A transformer that produces a higher voltage at its secondary than is applied to its primary.

switch A mechanical device used to connect or disconnect a component or circuit.

synchronous speed The speed of the rotating magnetic field of an AC induction motor.

temperature coefficient A ratio of the amount of change a temperature-sensing device makes when compared to the amount of change in temperature.

thermal compound A grease-like substance used to thermally bond two surfaces together for the purpose of increasing the rate of heat transfer from one object to another.

thermistor A device that changes its resistance with a change in temperature. These devices are made of metal oxides and exhibit a negative temperature coefficient.

thyristor An electronic component that has only two states of operation: on or off.

transistor A solid-state device made by combining three layers of semiconductor material together. A small amount of current flow through the base–emitter can control a larger amount of current flow through the collector–emitter.

triac A bidirectional thyristor device used to control AC voltage.

trigger pulse A voltage or current of short duration used to activate the gate, base, or input of some electronic device.

truth table A chart used to show the output condition of a logic gate or circuit as compared to different conditions of input.

UJT (unijunction transistor) A special transistor that is a member of the thyristor family of devices and operates like a voltage controlled switch.

unijunction transistor See *UJT.*

valence electron The electron in the outermost shell or orbit of an atom.

variable resistor A resistor whose resistance value can be adjusted between the limits of its minimum and maximum value.

volt/voltage An electrical measure of potential difference, electromotive force, or electrical pressure.

voltage divider A series connection of resistors used to produce different values of voltage drop across them.

voltage drop The amount of voltage required to cause an amount of current to flow through a certain value of resistance or reactance.

voltage follower A method of connection for an operational amplifier that will produce a gain of "1" but causes a great increase in input impedance.

voltage rating A rating that indicates the amount of voltage that can safely be connected to a device.

voltage regulator A device or circuit that maintains a constant value of voltage.

voltmeter An instrument used to measure a level of voltage.

volt-ohm-milliammeter (VOM) A test instrument so designed that it can be used to measure voltage, resistance, or milliamperes.

watt A measure of true power.

waveform The shape of a wave as obtained by plotting a graph with respect to voltage and time.

wye connection A connection of three components made in such a manner that one end of each component is connected. This connection is generally used to connect devices to a three-phase power system.

zener diode A special diode that exhibits a constant voltage drop when connected in such a manner that current flows through it in the reverse direction.

zener region The region current enters into when it flows through a diode in the reverse direction.

zero switching A feature of some solid-state relays such that current will continue to flow through the device until the AC waveform returns to zero.

INDEX